Springer Series in Statistics

Springer Series in Statistics

(continued after index)

Yves Tillé

Sampling Algorithms

 Springer

Yves Tillé
Institut de Statistique,
Université de Neuchâtel
Espace de l'Europe 4,
Case postale 805
2002 Neuchâtel,
Switzerland
yves.tille@unine.ch

Library of Congress Control Number: 2005937126

ISBN-10: 0-387-30814-8
ISBN-13: 978-0387-30814-2

Printed in the United States of America.   (MVY)

9 8 7 6 5 4 3 2 1

springer.com

Preface

This book is based upon courses on sampling algorithms. After having used scattered notes for several years, I have decided to completely rewrite the material in a consistent way. The books of Brewer and Hanif (1983) and Hájek (1981) have been my works of reference. Brewer and Hanif (1983) have drawn up an exhaustive list of sampling methods with unequal probabilities, which was probably a very tedious work. The posthumous book of Hájek (1981) contains an attempt at writing a general theory for conditional Poisson sampling. Since the publication of these books, things have been improving. New techniques of sampling have been proposed, to such an extent that it is difficult to have a general idea of the interest of each of them. I do not claim to give an exhaustive list of these new methods. Above all, I would like to propose a general framework in which it will be easier to compare existing methods. Furthermore, forty-six algorithms are precisely described, which allows the reader to easily implement the described methods.

This book is an opportunity to present a synthesis of my research and to develop my convictions on the question of sampling. At present, with the splitting method, it is possible to construct an infinite amount of new sampling methods with unequal probabilities. I am, however, convinced that conditional Poisson sampling is probably the best solution to the problem of sampling with unequal probabilities, although one can object that other procedures provide very similar results.

Another conviction is that the joint inclusion probabilities are not used for anything. I also advocate for the use of the cube method that allows selecting balanced samples. I would also like to apologize for all the techniques that are not cited in this book. For example, I do not mention all the methods called "order sampling" because the methods for coordinating samples are not examined in this book. They could be the topic of another publication.

This material is aimed at experienced statisticians who are familiar with the theory of survey sampling, to Ph.D. students who want to improve their knowledge in the theory of sampling and to practitioners who want to use or implement modern sampling procedures. The R package "sampling" available on the Web site of the Comprehensive R Archive Network (CRAN) contains an implementation of most of the described algorithms. I refer the reader to the books of Mittelhammer (1996) and Shao (2003) for questions of inferential statistics, and to the book of Särndal et al. (1992) for general questions related to the theory of sampling.

Finally, I would like to thank Jean-Claude Deville who taught me a lot on the topic of sampling when we worked together at the École Nationale de la Statistique et de l'Analyse de l'Information in Rennes from 1996 to 2000. I thank Yves-Alain Gerber, who has produced most of the figures of this book. I am also grateful to Cédric Béguin, Ken Brewer, Lionel Qualité, and Paul-André Salamin for their constructive comments on a previous version of this book. I am particularly indebted to Lennart Bondesson for his critical reading of the manuscript that allowed me to improve this book considerably and to Leon Jang for correction of the proofs.

Neuchâtel, October 2005

Yves Tillé

Contents

1

Introduction and Overview

1.1 Purpose

The aim of sampling methods is to provide information on a population by studying only a subset of it, called a sample. Sampling is the process of selecting units (households, businesses, people) from a population so that the sample allows estimating unknown quantities of the population. This book is an attempt to present a unified theory of sampling. The objective is also to describe precisely and rigorously the 46 procedures presented in this book. The R "sampling" package available on the R language Web site contains an implementation of most of the described algorithms.

1.2 Representativeness

One often says that a sample is representative if it is a reduced model of the population. Representativeness is then adduced as an argument of validity: a good sample must resemble the population of interest in such a way that some categories appear with the same proportions in the sample as in the population. This theory, currently spread by the media, is, however, erroneous. It is often more desirable to overrepresent some categories of the population or even to select units with unequal probabilities. The sample must not be a reduced model of the population; it is only a tool used to provide estimates.

Suppose that the aim is to estimate the production of iron in a country, and that we know that the iron is produced, on the one hand, by two huge steel companies with several thousands of workers and, on the other hand, by several hundreds of small craft-industries of less than 50 workers. Does the better design consist in selecting each unit with the same probability? Obviously, no. First, one will inquire about the production of the two biggest companies. Next, one will select a sample of small companies according to an appropriate sampling design. This simple example runs counter to the idea of representativeness and shows the interest of sampling with unequal probabilities.

Sampling with unequal probabilities is also very commonly used in the selfweighted two-stage sampling designs. The primary units are first selected with inclusion probabilities proportional to their sizes and next, the same number of secondary units is selected in each primary unit. This design has several advantages: the sample size is fixed and the work burden can easily be balanced between the interviewers.

Another definition of representativeness is that the sample must be random and all the statistical units must have a strictly positive probability of being selected. Otherwise, the sample has a problem of coverage and an unbiased estimation of some totals is impossible. This definition is obviously wise, but this term has been so overused that I think it should not be used anymore.

For Hájek (1981), a strategy is a pair consisting of a sampling design and an estimator. A strategy is said to be representative if it allows estimating a total of the population exactly, that is, without bias and with a null variance. If the simple Horvitz-Thompson estimator is used, a strategy can be representative only if the sample automatically reproduces some totals of the population; such samples are called balanced. A large part of this book is dedicated to balanced sampling.

1.3 The Origin of Sampling Theory

1.3.1 Sampling with Unequal Probabilities

Sampling methods were already mentioned in Tschuprow (1923) and Neyman (1934). Nevertheless, the first papers dedicated to sampling procedures were devoted to unequal probability sampling. Hansen and Hurwitz (1943) proposed the multinomial design with replacement with an unbiased estimator. Rapidly, it appears that unequal probability sampling without replacement is much more complicated. Numerous papers have been published, but most of the proposed methods are limited to a sample size equal to 2.

Brewer and Hanif (1983) constructed a very important synthesis in which 50 methods were presented according to their date of publication, but only 20 of them really work and many of the exact procedures are very slow to apply. The well-known systematic sampling design (see Madow, 1949; Sampford, 1967) procedure remain very good solutions for sampling with unequal probabilities.

1.3.2 Conditional Poisson Sampling

A very important method is Conditional Poisson Sampling (CPS), also called sampling with maximum entropy. This sampling design is obtained by selecting Poisson samples until a given sample size is obtained. Moreover, conditional Poisson sampling can also be obtained by maximizing the entropy subject to given inclusion probabilities. This sampling design appears to be natural as it maximizes the randomness of the selection of the sample.

Conditional Poisson sampling was really a Holy Grail for Hájek (1981), who dedicated a large part of his sampling research to this question. The main problem is that the inclusion probabilities of conditional Poisson sampling change according to the sample size. Hájek (1981) proposed several approximations of these inclusion probabilities in order to implement this sampling design. Chen et al. (1994) linked conditional Poisson sampling with the theory of exponential families, which has paved the way for several quick implementations of the method.

1.3.3 The Splitting Technique

Since the book of Brewer and Hanif (1983), several new methods of sampling with unequal probabilities have been published. In Deville and Tillé (1998), eight new methods were proposed, but the splitting method proposed in the same paper was also a way to present, sometimes more simply, almost all the existing methods. The splitting technique is thus a means to integrate the presentation of well-known methods and to make them comparable.

1.3.4 Balanced Sampling

The interest of balanced sampling was already pointed out more than 50 years ago by Yates (1946) and Thionet (1953). Several partial solutions of balanced sampling methods have been proposed by Deville et al. (1988), Ardilly (1991), Deville (1992), Hedayat and Majumdar (1995), and Valliant et al. (2000).

Royall and Herson (1973a,b) and Scott et al. (1978) discussed the importance of balanced sampling in order to protect the inference against a misspecified model. They proposed optimal estimators under a regression model. In this model-based approach, the optimality is conceived only with respect to the regression model without taking into account the sampling design. Nevertheless, these authors come to the conclusion that the sample must be balanced but not necessarily random. Curiously, they developed this theory while there still did not exist any general method for selecting balanced samples.

Recently, Deville and Tillé (2004) proposed a general procedure, the cube method, that allows the selection of random balanced samples on several balancing variables, with equal or unequal inclusion probabilities in the sense that the Horvitz-Thompson estimator is equal or almost equal to the population total of the balancing variables. Deville and Tillé (2005) have also proposed an approximation of the variance for the Horvitz-Thompson estimator in balanced sampling with large entropy.

1.4 Scope of Application

Sampling methods can be useful when it is impossible to examine all the units of a finite population. When a sampling frame or register is available, it is possible to obtain balancing information on the units of interest. In most cases, it is possible to take advantage of this information in order to increase the accuracy. In business statistics, registers are generally available. Several countries have a register of the population. Nevertheless, in the two-stage sampling design, the primary units are usually geographic areas, for which a large number of auxiliary variables are generally available.

The Institut National de la Statistique et de l'Analyse de l'Information (INSEE, France) has selected balanced samples for the most important statistical projects. In the redesigned census in France, a fifth of the municipalities with fewer than 10,000 inhabitants are sampled each year, so that after five years all the municipalities will be selected. All the households in these municipalities are surveyed. The five samples of municipalities are selected with equal probabilities using balanced sampling on a set of demographic variables. This methodology ensures the accuracy of the estimators based upon the five groups of rotation.

The selection of the primary units (areas) for INSEE's new master sample for household surveys is done with probabilities proportional to the number of dwellings using balanced sampling. Again, the balancing variables are a set of demographic and economic variables. The master sample is used over ten years to select all the households samples. This methodology ensures a better accuracy of the estimates depending on the master sample.

1.5 Aim of This Book

In view of the large number of publications, a synthesis is probably necessary. Recently, solutions have been given for two important problems in survey sampling: the implementation of conditional Poisson sampling and a rapid method for selecting balanced samples, but the research is never closed. A very open field is the problem of the co-ordination of samples in repeated surveys.

The challenge of this book is to propose at the same time a unified theory of the sampling methods and tools that are directly applicable to practical situations. The samples are formally defined as random vectors whose components denote the number of times a unit is selected in the sample. From the beginning, the notions of sampling design and sampling algorithm are differentiated. A sampling design is nothing more than a multivariate discrete distribution, whereas a sampling algorithm is a way of implementing a sampling design. Particular stress is given to a geometric representation of the sampling methods.

This book contains a very precise description of the sampling procedures in order to be directly applicable in real applications. Moreover, an implementation in the R language available on the website of the *Comprehensive R Archive Network* allows the reader to directly use the described methods (see Tillé and Matei, 2005).

1.6 Outline of This Book

In Chapter 2, a particular notation is defined. The sample is represented by a vector of $\mathbb{R}^N$ whose components are the number of times each unit is selected in the sample. The interest of this notation lies in the possibility of dealing with sampling in the same way with or without replacement. A sample therefore becomes a random vector with positive integer values and the sampling designs are discrete multivariate distributions.

In Chapter 3, general algorithms are presented. In fact, any particular sampling design can be implemented by means of several algorithms. Chapter 4 is devoted to simple random sampling. Again, an original definition is proposed. From a general expression, all the simple random sampling designs can be derived by means of change of support, a support being a set of samples.

Chapter 5 is dedicated to exponential sampling designs. Chen et al. (1994) have linked these sampling designs to random vectors with positive integer values and exponential distribution. Again, changes of support allow defining the most classic sampling designs, such as Poisson sampling, conditional Poisson sampling, and multinomial sampling. The exact implementation of the conditional Poisson design, that could also be called "exponential design with fixed sample size", was, since the book of Hájek, a Holy Grail that seemed inaccessible for reasons of combinatory explosion. Since the publication of the paper of Chen et al. (1994), important progress has been achieved in such a way that there now exist very fast algorithms that implement this design.

Chapter 6 is devoted to the splitting method proposed by Deville and Tillé (1998). This method is a process that allows constructing sampling procedures. Most of the sampling methods are best presented in the form of the splitting technique. A sequence of methods is presented, but it is possible to construct many other ones with the splitting method. Some well-known procedures, such as the Chao (1982) or the Brewer (1975) methods, are presented only by means of the splitting scheme because they are more comparable with this presentation.

In Chapter 7, are presented some sampling methods with unequal probabilities that are not exponential and that cannot be presented by means of the splitting scheme: systematic (see Madow, 1949) sampling, the Deville (1998) systematic method, and the Sampford (1967) method. The choice of the method is next discussed. When the same inclusion probabilities are used, it is possible to proof that none of the methods without replacement provides a better accuracy than the other ones. The choice of the method cannot thus be done in function of the accuracy, but essentially in function of practical considerations. The problem of variance estimation is also discussed. Two approximations are constructed by means of a simple sum of squares. Several sampling methods are compared to these approximations. One can notice that the variance of the Sampford method and the exponential procedure are very well adjusted by one of these approximations, which provides a guideline for the choice of a sampling method and of a variance estimator.

In Chapter 8, methods of balanced sampling are developed. After a presentation of the methods for equal inclusion probabilities, the cube method is described. This method allows selecting balanced samples on several dozens of variables. Two algorithms are presented, and the question of variance estimation in balanced sampling is developed. Finally, in Chapter 9, the cube method is applied in order to select a sample of municipalities in the Swiss canton of Ticino.

Population, Sample, and Sampling Design

2.1 Introduction

In this chapter, some basic concepts of survey sampling are introduced. An original notation is used. Indeed, a sample without replacement is usually defined as a subset of a finite population. However, in this book, we define a sample as a vector of indicator variables. Each component of this vector is the number of times that a unit is selected in the sample. This formalization is, however, not new. Among others, it has been used by Chen et al. (1994) for exponential designs and by Deville and Tillé (2004) for developing the cube method. This notation tends to be obvious because it allows defining a sampling design as a discrete multivariate distribution (Traat, 1997; Traat et al., 2004) and allows dealing with sampling with and without replacement in the same way. Of particular importance is the geometrical representation of a sampling design that is used in Chapter 8, which is dedicated to balanced sampling.

2.2 Population and Variable of Interest

A finite population is a set of N units $\{u_1, \ldots, u_k, \ldots, u_N\}$. Each unit can be identified without ambiguity by a label. Let

$$U = \{1, \ldots, k, \ldots, N\}$$

be the set of these labels. The size of the population N is not necessarily known.

The aim is to study a variable of interest y that takes the value y_k for unit k. Note that the y_k's are not random. The objective is more specifically to estimate a function of interest T of the y_k's:

$$T = f(y_1 \ \cdots \ y_k \ \cdots \ y_N).$$

The most common functions of interest are the total

$$Y = \sum_{k \in U} y_k,$$

the population size

$$N = \sum_{k \in U} 1,$$

the mean

$$\overline{Y} = \frac{1}{N} \sum_{k \in U} y_k,$$

the variance

$$\sigma_y^2 = \frac{1}{N} \sum_{k \in U} \left(y_k - \overline{Y}\right)^2 = \frac{1}{2N^2} \sum_{k \in U} \sum_{\ell \in U} (y_k - y_\ell)^2$$

and the corrected variance

$$V_y^2 = \frac{1}{N-1} \sum_{k \in U} \left(y_k - \overline{Y}\right)^2 = \frac{1}{2N(N-1)} \sum_{k \in U} \sum_{\ell \in U} (y_k - y_\ell)^2.$$

2.3 Sample

2.3.1 Sample Without Replacement

A sample without replacement is denoted by a column vector

$$\mathbf{s} = (s_1 \; \cdots \; s_k \; \cdots \; s_N)' \in \{0, 1\}^N,$$

where

$$s_k = \begin{cases} 1 & \text{if unit } k \text{ is in the sample} \\ 0 & \text{if unit } k \text{ is not in the sample,} \end{cases}$$

for all $k \in U$.

The sample size is

$$n(\mathbf{s}) = \sum_{k \in U} s_k.$$

Note that the null vector $\mathbf{s} = (0 \; \cdots \; 0 \; \cdots \; 0)'$ is a sample, called the empty sample and that the vector of ones $\mathbf{s} = (1 \; \cdots \; 1 \; \cdots \; 1)'$ is also a sample, called the census.

2.3.2 Sample With Replacement

The same kind of notation can be used for samples with replacement, which can also be denoted by a column vector

$$\mathbf{s} = (s_1 \cdots s_k \cdots s_N)' \in \mathbb{N}^N,$$

where $\mathbb{N} = \{0, 1, 2, 3, \dots\}$ is the set of natural numbers and s_k is the number of times that unit k is in the sample. Again, the sample size is

$$n(\mathbf{s}) = \sum_{k \in U} s_k,$$

and, in sampling with replacement, we can have $n(\mathbf{s}) > N$.

2.4 Support

Definition 1. *A support $\mathcal{Q}$ is a set of samples.*

Definition 2. *A support $\mathcal{Q}$ is said to be symmetric if, for any $\mathbf{s} \in \mathcal{Q}$, all the permutations of the coordinates of $\mathbf{s}$ are also in $\mathcal{Q}$.*

Some particular symmetric supports are used.

Definition 3. *The symmetric support without replacement is defined by*

$$\mathcal{S} = \{0, 1\}^N.$$

Note that $\mathrm{card}(\mathcal{S}) = 2^N$. The support $\mathcal{S}$ can be viewed as the set of all the vertices of a hypercube of $\mathbb{R}^N$ (or N-cube), where $\mathbb{R}$ is the set of real numbers, as shown for a population size $N = 3$ in Figure 2.1.

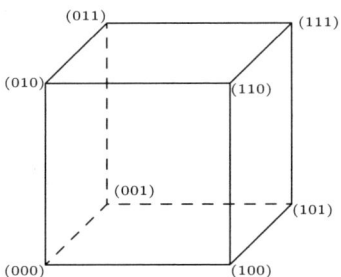

Fig. 2.1. Possible samples without replacement in a population of size $N = 3$

Definition 4. *The symmetric support without replacement with fixed sample size is*

$$\mathcal{S}_n = \left\{ \mathbf{s} \in \mathcal{S} \left| \sum_{k \in U} s_k = n \right. \right\}.$$

Note that $\text{card}(\mathcal{S}_n) = \binom{N}{n}$. Figure 2.2 shows an example of $\mathcal{S}_2$ in a population of size $N = 3$.

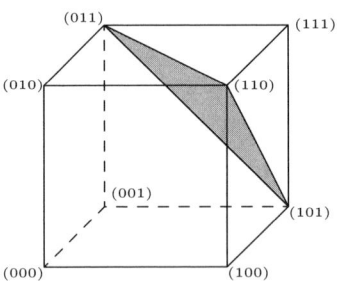

Fig. 2.2. The support $\mathcal{S}_2$ in a population of size $N = 3$

Definition 5. *The symmetric support with replacement is*

$$\mathcal{R} = \mathbb{N}^N,$$

where $\mathbb{N}$ is the set of natural numbers.

The full support with replacement $\mathcal{R}$ is countably infinite. As shown in Figure 2.3, support $\mathcal{R}$ is the lattice of all the vector of $\mathbb{N}^N$.

Definition 6. *The symmetric support with replacement of fixed size n is defined by*

$$\mathcal{R}_n = \left\{ \mathbf{s} \in \mathcal{R} \left| \sum_{k \in U} s_k = n \right. \right\}.$$

Result 1. *The size of $\mathcal{R}_n$ is*

$$\text{card}(\mathcal{R}_n) = \binom{N + n - 1}{n}. \tag{2.1}$$

Proof. (by induction) If $G(N, n)$ denotes the number of elements in $\mathcal{R}_n$ in a population of size N, then $G(1, n) = 1$. Moreover, we have the recurrence relation

$$G(N + 1, n) = \sum_{i=0}^{n} G(N, i). \tag{2.2}$$

Expression (2.1) satisfies the recurrence relation (2.2). $\square$

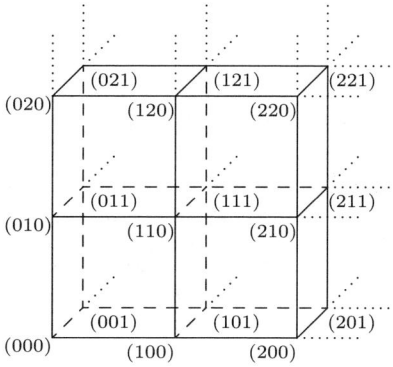

Fig. 2.3. Geometrical representation of support $\mathcal{R}$

Figure 2.4 shows the set of samples with replacement of size 2 in a population of size 3.

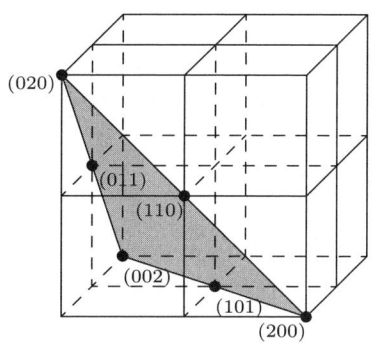

Fig. 2.4. Samples with replacement of size 2 in a population of size 3

The following properties follow directly:

1. $\mathcal{S}$, $\mathcal{S}_n$, $\mathcal{R}$, $\mathcal{R}_n$, are symmetric,
2. $\mathcal{S} \subset \mathcal{R}$,
3. The set $\{\mathcal{S}_0, \ldots, \mathcal{S}_n, \ldots, \mathcal{S}_N\}$ is a partition of $\mathcal{S}$,
4. The set $\{\mathcal{R}_0, \ldots, \mathcal{R}_n, \ldots, \mathcal{R}_N, \ldots\}$ is an infinite partition of $\mathcal{R}$,
5. $\mathcal{S}_n \subset \mathcal{R}_n$, for all $n = 0, \ldots, N$.

2.5 Convex Hull, Interior, and Subspaces Spanned by a Support

A support can be viewed as a set of points in $\mathbb{R}^N$. In what follows, several subsets of $\mathbb{R}^N$ are defined by means of the support: the convex hull, its interior, and several subspaces generated by the support. This geometrical vision is especially useful in Chapter 5 dedicated to exponential design and in Chapter 8 dedicated to balanced sampling. We refer to the following definitions.

Definition 7. *Let* $\mathbf{s}_1, \ldots, \mathbf{s}_i, \ldots, \mathbf{s}_I$ *be the enumeration of all the samples of* $\mathcal{Q}$. *The convex hull of the support* $\mathcal{Q}$ *is the set*

$$\mathrm{Conv}\,\mathcal{Q} = \left\{ \sum_{i=1}^{I} \lambda_i \mathbf{s}_i, \middle| \lambda_1, \ldots, \lambda_I, \text{of } \mathbb{R}_+, \text{such that } \sum_{i=1}^{I} \lambda_i = 1 \right\},$$

where $\mathbb{R}_+ = [0, \infty)$ *is the set of nonnegative real numbers.*

Definition 8. *Let* $\mathbf{s}_1, \ldots, \mathbf{s}_i, \ldots, \mathbf{s}_I$ *be the enumeration of all the samples of* $\mathcal{Q}$. *The affine subspace spanned by a support is the set*

$$\mathrm{Aff}\,\mathcal{Q} = \left\{ \sum_{i=1}^{I} \lambda_i \mathbf{s}_i, \middle| \lambda_1, \ldots, \lambda_I, \text{of } \mathbb{R}, \text{such that } \sum_{i=1}^{I} \lambda_i = 1 \right\}.$$

Definition 9. *The direction* $\overrightarrow{\mathcal{Q}}$ *of the affine subspace spanned by a support* $\mathcal{Q}$ *(the direction of the support for short) is the linear subspace spanned by all the differences* $\mathbf{s}_i - \mathbf{s}_j$, *for all* $\mathbf{s}_i, \mathbf{s}_j \in \mathcal{Q}$.

Note that $\overrightarrow{\mathcal{Q}}$ is a linear subspace.

Definition 10. *Let* $\mathbf{s}_1, \ldots, \mathbf{s}_i, \ldots, \mathbf{s}_I$ *be the enumeration of all the samples of* $\mathcal{Q}$. *The interior of* $\mathrm{Conv}\,\mathcal{Q}$ *in the affine subspace spanned by* $\mathcal{Q}$ *is defined by*

$$\overset{\circ}{\mathcal{Q}} = \left\{ \sum_{i=1}^{I} \lambda_i \mathbf{s}_i \middle| \lambda_i > 0, \text{ for all } i = 1, \ldots, I, \text{ and } \sum_{i=1}^{I} \lambda_i = 1 \right\}.$$

In short, afterwards $\overset{\circ}{\mathcal{Q}}$ *is called the interior of* $\mathcal{Q}$.

Definition 11. *The invariant subspace spanned by the support is defined by*

$$\mathrm{Invariant}\,\mathcal{Q} = \left\{ \mathbf{u} \in \mathbb{R}^N \middle| \mathbf{u}'(\mathbf{s}_1 - \mathbf{s}_2) = 0, \text{ for all } \mathbf{s}_1, \mathbf{s}_2 \in \mathcal{Q} \right\}.$$

Remark 1. $\overset{\circ}{\mathcal{Q}} \subset \mathrm{Conv}\,\mathcal{Q} \subset \mathrm{Aff}\,\mathcal{Q}$.

Remark 2. Invariant $\mathcal{Q}$ and $\overrightarrow{\mathcal{Q}}$ are two linear subspaces. Invariant $\mathcal{Q}$ is the orthogonal subspace of $\overrightarrow{\mathcal{Q}}$.

Table 2.1 gives the list of the most common supports and their invariant subspace, direction, affine subspace, convex hull, and interior, where $\mathbb{R}_+^* = \{x \in \mathbb{R} | x > 0\}$ is the set of positive real numbers.

Table 2.1. Invariant, direction, affine subspace, convex hull, and interior of the supports $\mathcal{R}, \mathcal{R}_n, \mathcal{S},$ and $\mathcal{S}_n$

	$\mathcal{R}$	$\mathcal{R}_n$	$\mathcal{S}$	$\mathcal{S}_n$
Invariant	$\{\mathbf{0}\}$	$\left\{\mathbf{x} \in \mathbb{R}^N \middle| \mathbf{x} = a\mathbf{1}, \text{ for all } a \in \mathbb{R}\right\}$	$\{\mathbf{0}\}$	$\left\{\mathbf{x} \in \mathbb{R}^N \middle| \mathbf{x} = a\mathbf{1}, \text{ for all } a \in \mathbb{R}\right\}$
Direction	$\mathbb{R}^N$	$\left\{\mathbf{u} \in \mathbb{R}^N \middle| \sum_{k=1}^N u_i = 0\right\}$	$\mathbb{R}^N$	$\left\{\mathbf{u} \in \mathbb{R}^N \middle| \sum_{k=1}^N u_i = 0\right\}$
Aff	$\mathbb{R}^N$	$\left\{\mathbf{v} \in \mathbb{R}^N \middle| \sum_{i=1}^N v_i = n\right\}$	$\mathbb{R}^N$	$\left\{\mathbf{v} \in \mathbb{R}^N \middle| \sum_{i=1}^N v_i = n\right\}$
Conv	$(\mathbb{R}_+)^N$	$\left\{\mathbf{v} \in [0,n]^N \middle| \sum_{i=1}^N v_i = n\right\}$	$[0,1]^N$	$\left\{\mathbf{v} \in [0,1]^N \middle| \sum_{i=1}^N v_i = n\right\}$
Interior	$(\mathbb{R}_+^*)^N$	$\left\{\mathbf{v} \in]0,n[^N \middle| \sum_{i=1}^N v_i = n\right\}$	$]0,1[^N$	$\left\{\mathbf{v} \in]0,1[^N \middle| \sum_{i=1}^N v_i = n\right\}$

Example 1. Let $U = \{1,2,3,4\}$ and

$$\mathcal{Q} = \left\{ \begin{pmatrix} 1 \\ 0 \\ 1 \\ 0 \end{pmatrix}, \begin{pmatrix} 1 \\ 0 \\ 0 \\ 1 \end{pmatrix}, \begin{pmatrix} 0 \\ 1 \\ 1 \\ 0 \end{pmatrix}, \begin{pmatrix} 0 \\ 1 \\ 0 \\ 1 \end{pmatrix} \right\}.$$

Then

$$\text{Aff } \mathcal{Q} = \left\{ \mathbf{u} \in \mathbb{R}^N \middle| u_1 + u_2 = 1 \text{ and } u_3 + u_4 = 1 \right\},$$

$$\text{Conv } \mathcal{Q} = \left\{ \mathbf{u} \in [0,1]^N \middle| u_1 + u_2 = 1 \text{ and } u_3 + u_4 = 1 \right\},$$

$$\overset{\circ}{\mathcal{Q}} = \left\{ \mathbf{u} \in]0,1[^N \middle| u_1 + u_2 = 1 \text{ and } u_3 + u_4 = 1 \right\},$$

$$\text{Invariant } \mathcal{Q} = \left\{ a \begin{pmatrix} 1 \\ 1 \\ 0 \\ 0 \end{pmatrix} + b \begin{pmatrix} 0 \\ 0 \\ 1 \\ 1 \end{pmatrix}, \text{ for all } a,b \in \mathbb{R} \right\},$$

and

$$\overrightarrow{\mathcal{Q}} = \left\{ a \begin{pmatrix} 1 \\ -1 \\ 0 \\ 0 \end{pmatrix} + b \begin{pmatrix} 0 \\ 0 \\ 1 \\ -1 \end{pmatrix}, \text{ for all } a,b \in \mathbb{R} \right\}.$$

2.6 Sampling Design and Random Sample

Definition 12. *A sampling design $p(.)$ on a support $\mathcal{Q}$ is a multivariate probability distribution on $\mathcal{Q}$; that is, $p(.)$ is a function from support $\mathcal{Q}$ to $]0,1]$ such that $p(\mathbf{s}) > 0$ for all $\mathbf{s} \in \mathcal{Q}$ and*

$$\sum_{\mathbf{s} \in \mathcal{Q}} p(\mathbf{s}) = 1.$$

Definition 13. *A sampling design with support $\mathcal{Q}$ is said to be without replacement if $\mathcal{Q} \subset \mathcal{S}$.*

Definition 14. *A sampling design with support $\mathcal{Q}$ is said to be of fixed sample size n if $\mathcal{Q} \subset \mathcal{R}_n$.*

Remark 3. Because $\mathcal{S}$ can be viewed as the set of all the vertices of a hypercube, a sampling design without replacement is a probability measure on all these vertices.

Definition 15. *A sampling design of support $\mathcal{Q}$ is said to be with replacement if $\mathcal{Q} \backslash \mathcal{S}$ is nonempty.*

Definition 16. *If $p(.)$ is a sampling design without replacement, the complementary design $p^c(.)$ of a sampling design $p(c)$ of support $\mathcal{Q}$ is*

$$p^c(\mathbf{1} - \mathbf{s}) = p(\mathbf{s}), \ \ for \ all \ \mathbf{s} \in \mathcal{Q},$$

where $\mathbf{1}$ is the vector of ones of $\mathbb{R}^N$.

Definition 17. *A random sample $\mathbf{S} \in \mathbb{R}^N$ with the sampling design $p(.)$ is a random vector such that*

$$\Pr(\mathbf{S} = \mathbf{s}) = p(\mathbf{s}), \ \ for \ all \ \mathbf{s} \in \mathcal{Q},$$

where $\mathcal{Q}$ is the support of $p(.)$.

2.7 Reduction of a Sampling Design With Replacement

Let $\mathbf{S} = (S_1 \cdots S_k \cdots S_N)'$ be a sample with replacement. The reduction function $r(.)$ suppresses the information about the multiplicity of the units, as follows. If

$$\mathbf{S}^* = r(\mathbf{S}),$$

then

$$S_k^* = \begin{cases} 1 & \text{if } S_k > 0 \\ 0 & \text{if } S_k = 0. \end{cases} \tag{2.3}$$

The reduction function allows constructing a design without replacement from a design with replacement. If $p(.)$ is a sampling design with support $\mathcal{R}$, then it is possible to derive a design $p^*(.)$ on $\mathcal{S}$ in the following way:

$$p^*(\mathbf{s}^*) = \sum_{\mathbf{s} \in \mathcal{R} \mid \mathbf{s}^* = r(\mathbf{s})} p(\mathbf{s}),$$

for all $\mathbf{s} \in \mathcal{S}$.

Example 2. Let $p(.)$ be a sampling design with replacement of fixed size $n = 2$ on $U = \{1, 2, 3\}$ such that

$$p((2,0,0)') = \frac{1}{9}, \quad p((0,2,0)') = \frac{1}{9}, \quad p((0,0,2)') = \frac{1}{9},$$

$$p((1,1,0)') = \frac{2}{9}, \quad p((1,0,1)') = \frac{2}{9}, \quad p((0,1,1)') = \frac{2}{9};$$

then $p^*(.)$ is given by

$$p((1,0,0)') = \frac{1}{9}, \quad p((0,1,0)') = \frac{1}{9}, \quad p((0,0,1)') = \frac{1}{9},$$

$$p((1,1,0)') = \frac{2}{9}, \quad p((1,0,1)') = \frac{2}{9}, \quad p((0,1,1)') = \frac{2}{9}.$$

2.8 Expectation and Variance of a Random Sample

Definition 18. *The expectation of a random sample* $\mathbf{S}$ *is*

$$\boldsymbol{\mu} = \mathrm{E}(\mathbf{S}) = \sum_{\mathbf{s} \in \mathcal{Q}} p(\mathbf{s})\mathbf{s}.$$

Remark 4. Because $\boldsymbol{\mu}$ is a linear convex combination with positive coefficient of the elements of $\mathcal{Q}$, $\boldsymbol{\mu} \in \overset{\circ}{\mathcal{Q}}$.

Remark 5. Let $\mathbf{C}$ be the $N \times I$ matrix of all the centered samples of $\mathcal{Q}$; that is

$$\mathbf{C} = (\mathbf{s}_1 - \boldsymbol{\mu} \;\; \cdots \;\; \mathbf{s}_I - \boldsymbol{\mu}), \tag{2.4}$$

where $I = \mathrm{card}\,\mathcal{Q}$. Then $\mathrm{Ker}\,\mathbf{C}' = \mathrm{Invariant}\,\mathcal{Q}$, where

$$\mathrm{Ker}\,\mathbf{C}' = \{\mathbf{u} \in \mathbb{R}^N \mid \mathbf{C}'\mathbf{u} = \mathbf{0}\}.$$

The joint expectation is defined by

$$\mu_{k\ell} = \sum_{\mathbf{s} \in \mathcal{Q}} p(\mathbf{s})s_k s_\ell.$$

Finally, the variance-covariance operator is

$$\boldsymbol{\Sigma} = [\Sigma_{k\ell}] = \mathrm{var}(\mathbf{S}) = \sum_{\mathbf{s} \in \mathcal{Q}} p(\mathbf{s})(\mathbf{s} - \boldsymbol{\mu})(\mathbf{s} - \boldsymbol{\mu})' = [\mu_{k\ell} - \mu_k \mu_\ell].$$

Result 2.
$$\sum_{k \in U} \mu_k = \mathrm{E}[n(\mathbf{S})],$$

and, if the sampling design is of fixed sample size $n(\mathbf{S})$,

$$\sum_{k \in U} \mu_k = n(\mathbf{S}).$$

Proof. Because

$$\sum_{k \in U} S_k = n(\mathbf{S}),$$

we have

$$\sum_{k \in U} \mathrm{E}(S_k) = \sum_{k \in U} \mu_k = \mathrm{E}\left[n(\mathbf{S})\right]. \qquad \square$$

Result 3. *Let* $\boldsymbol{\Sigma} = [\varSigma_{k\ell}]$ *be the variance-covariance operator. Then*

$$\sum_{k \in U} \varSigma_{k\ell} = \mathrm{E}\left[n(\mathbf{S})(S_\ell - \mu_\ell)\right], \text{ for all } \ell \in U,$$

and, if the sampling design is of fixed sample size n,

$$\sum_{k \in U} \varSigma_{k\ell} = 0, \text{ for all } \ell \in U. \tag{2.5}$$

Proof. Because

$$\sum_{k \in U} \varSigma_{k\ell} = \sum_{k \in U} (\mu_{k\ell} - \mu_k \mu_\ell) = \sum_{k \in U} [\mathrm{E}(S_k S_\ell) - \mathrm{E}(S_k)\mathrm{E}(S_\ell)]$$
$$= \mathrm{E}\left[n(\mathbf{S})(S_\ell - \mu_\ell)\right].$$

If $\mathrm{var}[n(\mathbf{S})] = 0$, then

$$\sum_{k \in U} \varSigma_{k\ell} = \mathrm{E}\left[n(\mathbf{S})(S_\ell - \mu_\ell)\right] = n(\mathbf{S})\mathrm{E}\left(S_\ell - \mu_\ell\right) = 0. \qquad \square$$

Result 4. *Let* $p(.)$ *be a sampling design with support* $\mathcal{Q}$. *Then*

$$\mathrm{Ker}\ \boldsymbol{\Sigma} = \mathrm{Invariant}\ \mathcal{Q}.$$

Proof. Let $\mathbf{C}$ denote the matrix of the centered samples defined in (2.4). Because $\boldsymbol{\Sigma} = \mathbf{CPC}'$ where $\mathbf{P} = \mathrm{diag}[p(\mathbf{s}_1) \ \cdots \ p(\mathbf{s}_I)]$, we have that, if $\mathbf{u} \in$ Invariant $\mathcal{Q}$, $\mathbf{C}'\mathbf{u} = \mathbf{0}$, thus $\boldsymbol{\Sigma}\mathbf{u} = \mathbf{0}$; that is, Ker $\boldsymbol{\Sigma} \subset$ Invariant $\mathcal{Q}$. Moreover, because $\mathbf{P}$ is of full rank, $\mathbf{C}$ and $\boldsymbol{\Sigma}$ have the same rank, which implies that Ker $\boldsymbol{\Sigma}$ and Ker $\mathbf{C}$ have the same dimension. Thus, Ker $\boldsymbol{\Sigma} =$ Invariant $\mathcal{Q}$. $\quad\square$

Remark 6. Let the image of $\boldsymbol{\Sigma}$ be

$$\text{Im}(\boldsymbol{\Sigma}) = \left\{ \mathbf{x} \in \mathbb{R}^N \,\middle|\, \text{there exists } \mathbf{u} \in \mathbb{R}^N \text{ such that } \boldsymbol{\Sigma}\mathbf{u} = \mathbf{x} \right\}.$$

Because $\text{Im}(\boldsymbol{\Sigma})$ is the orthogonal of $\text{Ker}\boldsymbol{\Sigma}$, $\text{Im}(\boldsymbol{\Sigma}) = \vec{\mathcal{Q}}$.

Example 3. If the sample is of fixed sample size, $\mathbf{1} \in$ Invariant $\mathcal{Q}$, $\boldsymbol{\Sigma}\mathbf{1} = \mathbf{0}$ where $\mathbf{1} = (1 \;\cdots\; 1 \;\cdots\; 1)'$. Result 4 is thus a generalization of Expression (2.5).

2.9 Inclusion Probabilities

Definition 19. *The first-order inclusion probability is the probability that unit k is in the random sample*

$$\pi_k = \Pr(S_k > 0) = \text{E}[r(S_k)],$$

where $r(.)$ is the reduction function.

Moreover, $\boldsymbol{\pi} = (\pi_1 \;\cdots\; \pi_k \;\cdots\; \pi_N)'$ denotes the vector of inclusion probabilities. If the sampling design is without replacement, then $\boldsymbol{\pi} = \boldsymbol{\mu}$.

Definition 20. *The joint inclusion probability is the probability that unit k and ℓ are together in the random sample*

$$\pi_{k\ell} = \Pr(S_k > 0 \text{ and } S_\ell > 0) = \text{E}\left[r(S_k)r(S_\ell)\right],$$

with $\pi_{kk} = \pi_k, k \in U$.

Let $\boldsymbol{\Pi} = [\pi_{k\ell}]$ be the matrix of joint inclusion probabilities. Moreover, we define

$$\boldsymbol{\Delta} = \boldsymbol{\Pi} - \boldsymbol{\pi}\boldsymbol{\pi}'.$$

Note that $\Delta_{kk} = \pi_k(1 - \pi_k), k \in U$. If the sample is without replacement, then

$$\boldsymbol{\Delta} = \boldsymbol{\Sigma}.$$

Result 5.
$$\sum_{k \in U} \pi_k = \text{E}\left\{n[r(\mathbf{S})]\right\},$$

and

$$\sum_{k \in U} \Delta_{k\ell} = \text{E}\left\{n[r(\mathbf{S})]\left(r(S_\ell) - \pi_\ell\right)\right\}, \text{ for all } \ell \in U.$$

Moreover, if $\text{var}\left\{n[r(\mathbf{S})]\right\} = 0$ *then*

$$\sum_{k \in U} \Delta_{k\ell} = 0, \text{ for all } \ell \in U.$$

The proof is the same as for Results 2 and 3.

When the sample is without replacement and of fixed sample size, then $\mathbf{S} = r(\mathbf{S})$, and Result 5 becomes

$$\sum_{k \in U} \pi_k = n(\mathbf{S}),$$

and

$$\sum_{k \in U} \Delta_{k\ell} = 0, \text{ for all } \ell \in U.$$

2.10 Computation of the Inclusion Probabilities

Suppose that the values x_k of an auxiliary variable x are known for all the units of U and that the $x_k > 0$, for all $k \in U$. When the x_k's are approximately proportional to the values y_k's of the variable of interest, it is then interesting to select the units with unequal probabilities proportional to the x_k in order to get an estimator of Y with a small variance. Unequal probability sampling is also used in multistage selfweighted sampling designs (see, for instance, Särndal et al., 1992, p. 141).

To implement such a sampling design, the inclusion probabilities π_k are computed as follows. First, compute the quantities

$$\frac{n x_k}{\sum_{\ell \in U} x_\ell}, \tag{2.6}$$

$k = 1, \ldots, N$. For units for which these quantities are larger than 1, set $\pi_k = 1$. Next, the quantities are recalculated using (2.6) restricted to the remaining units. This procedure is repeated until each π_k is in $]0, 1]$. Some π_k are 1 and the others are proportional to x_k. This procedure is formalized in Algorithm 2.1.

More formally, define

$$h(z) = \sum_{k \in U} \min \left(z \frac{x_k}{X}, 1 \right), \tag{2.7}$$

where $X = \sum_{k \in U} x_k$. Then, the inclusion probabilities are given by

$$\pi_k = \min \left[1, h^{-1}(n) \frac{x_k}{X} \right], k \in U. \tag{2.8}$$

The selection of the units with $\pi_k = 1$ is trivial, therefore we can consider that the problem consists of selecting n units without replacement from a population of size N with fixed π_k, where $0 < \pi_k < 1, k \in U$, and

$$\sum_{k \in U} \pi_k = n.$$

Algorithm 2.1 Computation of the inclusion probabilities

DEFINITION **f**: VECTORS OF N INTEGERS;
   $\boldsymbol{\pi}$: VECTORS OF N REALS;
   X REAL,
   $m, n, m1$ INTEGER;
FOR $k = 1, \ldots, N$ DO $f_k = 0$; ENDFOR;
$m = 0$;
$m1 = -1$;
WHILE $(m1 \neq m)$, DO
  $m1 = m$;
  $X = \sum_{k=1}^{N} x_k(1 - f_k)$;
  $c = n - m$;
  FOR $k = 1, \ldots, N$ DO
   IF $f_k = 0$ THEN $\pi_k = cx_k/X$; IF $\pi_k \geq 1$ THEN $f_k = 1$; ENDIF;
   ELSE $\pi_k = 1$;
   ENDIF;
  ENDFOR;
  $m = \sum_{k=1}^{N} f_k$;
ENDWHILE.

2.11 Characteristic Function of a Sampling Design

A sampling design can thus be viewed as a multivariate distribution, which allows defining the characteristic function of a random sample.

Definition 21. *The characteristic function $\phi(\mathbf{t})$ from $\mathbb{R}^N$ to $\mathbb{C}$ of a random sample $\mathbf{S}$ with sampling design $p(.)$ on $\mathcal{Q}$ is defined by*

$$\phi_{\mathbf{S}}(\mathbf{t}) = \sum_{\mathbf{s} \in \mathcal{Q}} e^{i\mathbf{t}'\mathbf{s}} p(\mathbf{s}), \mathbf{t} \in \mathbb{R}^N, \tag{2.9}$$

where $i = \sqrt{-1}$, and $\mathbb{C}$ is the set of the complex numbers.

In some cases (see Chapters 4 and 5), the expression of the characteristic function can be significantly simplified in such a way that the sum over all the samples of the support does not appear anymore. When the characteristic function can be simplified, the sampling design is more tractable.

 If $\phi'(\mathbf{t})$ denotes the vector of first derivatives and $\phi''(\mathbf{t})$ the matrix of second derivatives of $\phi(\mathbf{t})$, then we have

$$\phi'(0) = i \sum_{\mathbf{s} \in \mathcal{Q}} \mathbf{s} p(\mathbf{s}) = i\boldsymbol{\mu},$$

and

$$\phi''(0) = -\sum_{\mathbf{s} \in \mathcal{Q}} \mathbf{s}\mathbf{s}' p(\mathbf{s}) = -(\boldsymbol{\Sigma} + \boldsymbol{\mu}\boldsymbol{\mu}').$$

2.12 Conditioning a Sampling Design

A sampling design can be conditioned with respect to a support.

Definition 22. *Let $\mathcal{Q}_1$ and $\mathcal{Q}_2$ be two supports such that $\mathcal{Q}_2 \subset \mathcal{Q}_1$ and $p_1(.)$ a sampling design on $\mathcal{Q}_1$. Then, the conditional design of $\mathcal{Q}_1$ with respect to $\mathcal{Q}_2$ is given by*

$$p_1(\mathbf{s}|\mathcal{Q}_2) = \frac{p_1(\mathbf{s})}{\sum_{\mathbf{s} \in \mathcal{Q}_2} p_1(\mathbf{s})}, \quad \text{for all } \mathbf{s} \in \mathcal{Q}_2.$$

A sampling design $p(.)$ can also be conditioned with respect to a particular value a_k for S_k

$$p(\mathbf{s}|S_k = a_k) = \frac{p(\mathbf{s})}{\sum_{\mathbf{s}|s_k=a_k} p(\mathbf{s})},$$

for all $\mathbf{s}$ such that $S_k = a_k$.

Finally, a sampling design can also be conditioned with respect to particular values a_k for a subset $A \subset U$ of units in the sample

$$p\left[\mathbf{s} \,\middle|\, \bigcap_{k \in A}(S_k = a_k)\right] = \frac{p(\mathbf{s})}{\sum_{\mathbf{s}|\bigcap_{k \in A}(s_k=a_k)} p(\mathbf{s})}.$$

for all $\mathbf{s}$ such that $s_k = a_k, k \in A$.

2.13 Observed Data and Consistency

Definition 23. *The set $D = \{(V_k, S_k), k \in U\}$ is called the observed data, where $V_k = y_k S_k$.*

Also, let d denote a possible value for D

$$d = \{(v_k, s_k), k \in U\},$$

where $v_k = y_k s_k$. The set of possible values for the observed data is

$$\mathcal{X} = \{d | \mathbf{s} \in \mathcal{S}, \mathbf{y} \in \mathbb{R}^N\}.$$

The following definition is necessary to define the probability distribution of D.

Definition 24. *A possible value d is said to be consistent with a particular vector $\mathbf{y}_N^* = (y_1^* \cdots y_k^* \cdots y_N^*)$ if and only if $y_k = y_k^*$, for all k such that $s_k > 0$.*

Because

$$\Pr(D = d; \mathbf{y}) = \Pr(D = d \text{ and } \mathbf{S} = \mathbf{s}; \mathbf{y}) = \Pr(D = d | \mathbf{S} = \mathbf{s}; \mathbf{y})\Pr(\mathbf{S} = \mathbf{s}),$$

$$\Pr(D = d | \mathbf{S} = \mathbf{s}; \mathbf{y}) = \begin{cases} 1 & \text{if } d \text{ is consistent with } \mathbf{y} \\ 0 & \text{otherwise.} \end{cases}$$

The probability distribution of D can be written

$$p_Y(d) = \Pr(D = d; \mathbf{y}) = \begin{cases} p(\mathbf{s}) & \text{if } d \text{ is consistent with } \mathbf{y} \\ 0 & \text{otherwise.} \end{cases} \tag{2.10}$$

2.14 Statistic and Estimation

Definition 25. *A statistic G is a function of the observed data: $G = u(D)$.*

The expectation of G is

$$E(G) = \sum_{\mathbf{s} \in \mathcal{Q}} p(\mathbf{s})G(\mathbf{s}),$$

where $G(\mathbf{s})$ is the value taken by statistic G on sample $\mathbf{s}$, and $\mathcal{Q}$ is the support of the sampling design. The variance of a statistic G is defined by

$$\text{var}(G) = E\left[G - E(G)\right]^2.$$

Definition 26. *An estimator $\widehat{T}$ of a function of interest T is a statistic used to estimate it.*

The bias of $\widehat{T}$ is defined by

$$B(\widehat{T}) = E(\widehat{T}) - T,$$

and the mean square error by

$$\text{MSE}(\widehat{T}) = E\left(\widehat{T} - T\right)^2 = \text{var}(\widehat{T}) + B^2(\widehat{T}).$$

Definition 27. *An estimator $\widehat{Y}_L$ of Y is said to be linear if it can be written*

$$\widehat{Y}_{L1} = w_0 + \sum_{k \in U} w_k y_k S_k,$$

or

$$\widehat{Y}_{L2} = w_0 + \sum_{k \in U} w_k y_k r(S_k),$$

where w_k can eventually depend on $\mathbf{S}$ and thus can be random.

Result 6. *Let $p(.)$ be a sampling design such that $\mu_k > 0$, for all $k \in U$. A linear estimator $\widehat{Y}_{L1}$ of Y is unbiased if and only if*

$$\mathrm{E}(w_0) = 0 \quad \text{and} \quad \mathrm{E}(w_k S_k) = 1, \ \text{for all } k \in U.$$

A linear estimator $\widehat{Y}_{L2}$ of Y is unbiased if and only if

$$\mathrm{E}(w_0) = 0 \quad \text{and} \quad \mathrm{E}[w_k r(S_k)] = 1, \ \text{for all } k \in U.$$

Because the y_k's are not random, the proof is obvious.

2.15 Sufficient Statistic

In order to deal with the estimation problem, one could use the technique of sufficiency reduction. We refer to the following definition.

Definition 28. *A statistic $G = u(D)$ is said to be sufficient for vector $\mathbf{y} = (y_1 \cdots y_N)'$ if and only if the conditional distribution of D given that $G = g$ does not depend on $\mathbf{y} = (y_1 \cdots y_N)'$ (as long as this conditional probability exists).*

If $\Pr(D = d | G = g; \mathbf{y})$ is the conditional probability, a statistic G is sufficient if $\Pr(D = d | G = g; \mathbf{y})$ is constant with respect to $\mathbf{y}$ as long as $\Pr(G = g; \mathbf{y}) > 0$.

A sample with replacement can always be reduced to a sample without replacement of $\mathcal{S}$ by means of the reduction function $r(.)$ given in Expression (2.3) that suppresses the information about the multiplicity of the units. A sampling design $p(.)$ with support $\mathcal{Q} \subset \mathcal{R}$ can also be reduced to a sampling design $p^*(.)$ without replacement:

$$p^*(\mathbf{s}^*) = \sum_{\mathbf{s} \in \mathcal{Q} | r(\mathbf{s}) = \mathbf{s}^*} p(\mathbf{s}).$$

Similarly, if the sample is with replacement, the observed data can also be reduced using function $u^*(.)$, where

$$u^*(D) = \left\{ \left(\frac{V_k}{1 - r(S_k) + S_k}, r(S_k) \right), k \in U \right\} = \{(y_k r(S_k), r(S_k)), k \in U\}$$

and $V_k = y_k S_k$.

Theorem 1 (*see Basu and Ghosh, 1967; Basu, 1969; Cassel et al., 1993, p. 36; Thompson and Seber, 1996, p. 35*). *For all sampling designs with support $\mathcal{Q} \subset \mathcal{R}$, $D^* = u^*(D)$ is a sufficient statistic for $\mathbf{y}$.*

Proof. Consider the conditional probability:

$$\Pr(D = d | D^* = d^*; \mathbf{y}) = \frac{\Pr(D = d \text{ and } D^* = d^*; \mathbf{y})}{\Pr(D^* = d^*; \mathbf{y})}.$$

This probability is defined only if $\Pr(D^* = d^*; \mathbf{y}) > 0$, which implies that D^* is consistent with $\mathbf{y}$.

Two cases must be distinguished.

1. If $u^*(d) = d^*$, then by (2.10) we obtain

$$\Pr(D = d \text{ and } D^* = d^*; \mathbf{y}) = \Pr(D = d; \mathbf{y}) = p_Y(d) = p(\mathbf{s}).$$

2. If $u^*(d) \neq d^*$, then $\Pr(D = d \text{ and } D^* = d^*; \mathbf{y}) = 0$.

If d^* is consistent with $\mathbf{y}$, we have

$$\Pr(D = d; \mathbf{y}) = p(\mathbf{s}),$$

and thus

$$\Pr(D = d | D^* = d^*; \mathbf{y}) = \begin{cases} p(\mathbf{s})/p^*(\mathbf{s}^*) & \text{if } d^* = u^*(d) \\ 0 & \text{if } d^* \neq u^*(d), \end{cases}$$

where $r(\mathbf{s}) = \mathbf{s}^*$ and

$$p^*(\mathbf{s}^*) = \sum_{\mathbf{s} \in \mathcal{Q} | r(\mathbf{s}) = \mathbf{s}^*} p(\mathbf{s}).$$

This conditional distribution does not depend on $\mathbf{y}$ and $u^*(D)$ is thus sufficient. □

The factorization theorem allows for the identification of a sufficient statistic:

Theorem 2 (*see Basu and Ghosh, 1967; Basu, 1969; Cassel et al., 1993, p. 36*). *For all sampling designs on $\mathcal{Q} \subset \mathcal{R}$ then $G = u(D)$ is sufficient for $\mathbf{y}$ if and only if*

$$p_Y(d) = g[u(d), \mathbf{y}]h(d),$$

where the function $h(d)$ does not depend on $\mathbf{y}$ and $g(.)$ only depends on $u(d)$.

Proof. Define the indicator function

$$\delta(d, \mathbf{y}) = \begin{cases} 1 & \text{if } d \text{ is consistent with } \mathbf{y} \\ 0 & \text{otherwise.} \end{cases}$$

Expression (2.10) can thus be written:

$$p_Y(d) = p(\mathbf{s})\delta(d; \mathbf{y}). \tag{2.11}$$

If $G = u(D)$ is a sufficient statistic and $\Pr[G = u(d); \mathbf{y}] > 0$, we have

$$\Pr[D = d | G = u(d); \mathbf{y}] = h(d),$$

where $h(d)$ does not depend on $\mathbf{y}$. We thus have

$$\begin{aligned} p_Y(d) &= \Pr(D = d \text{ and } G = u(d); \mathbf{y}) \\ &= \Pr(G = u(d); \mathbf{y})h(d) \\ &= g[u(d), \mathbf{y}]h(d), \end{aligned}$$

where $g[u(d), \mathbf{y}]$ only depends on $u(d)$. □

The reduction principle consists of retaining from the observed data a minimum of information in the following way:

Definition 29. *A statistic is said to be a minimum sufficient statistic if it can be written as a function of all the sufficient statistics.*

Theorem 3 (*see Basu and Ghosh, 1967; Basu, 1969; Cassel et al., 1993, p. 37; Thompson and Seber, 1996, p. 38*). *For all sampling designs with replacement, the statistic $D^* = u^*(D)$ (suppression of the information about the multiplicity of the units) is minimum sufficient.*

Proof. A statistic $u(.)$ can define a partition of $\mathcal{X}$ denoted $\mathcal{P}_u$. This partition is such that if $u(d_1) = u(d_2)$ then d_1 and d_2 are in the same subset of the partition. A statistic $u^*(.)$ is thus minimum sufficient if, for every sufficient statistic $u(.)$, each subset of the partition $\mathcal{P}_u$ is included in a subset of the partition $\mathcal{P}_{u^*}$.

Let $u(.)$ be any sufficient statistic such that $u(d_1) = u(d_2)$, which means that d_1 and d_2 are in the same subset of the partition $\mathcal{P}_u$. We show that it implies that they are in the same subset of the partition $\mathcal{P}_{u^*}$. Because $u(.)$ is sufficient, by Theorem 2, we have

$$p_Y(d_1) = g\left[u(d_1), \mathbf{y}\right] h(d_1)$$

and

$$p_Y(d_2) = g\left[u(d_2), \mathbf{y}\right] h(d_2).$$

Because $g[u(d_1), \mathbf{y}] = g[u(d_2), \mathbf{y}]$ and $u(d_1) = u(d_2)$, we obtain

$$\frac{p_Y(d_1)}{h(d_1)} = \frac{p_Y(d_2)}{h(d_2)}.$$

Finally, by (2.11), we have

$$\frac{p(\mathbf{s}_1)\delta(d_1; \mathbf{y})}{h(d_1)} = \frac{p(\mathbf{s}_2)\delta(d_2; \mathbf{y})}{h(d_2)},$$

which is true if and only if $\delta(d_1; \mathbf{y}) = \delta(d_2; \mathbf{y})$. In this case, $r(\mathbf{s}_1) = r(\mathbf{s}_2)$ and $u^*(d_1) = u^*(d_2)$. Thus, d_1 and d_2 are in the same subset of the partition $\mathcal{P}_{u^*}$. Because this result is true for any sufficient statistic $u(.)$, the statistic $u^*(.)$ is minimum sufficient. □

This result shows that the information about the multiplicity of the units can always be suppressed from the observed data. The Rao-Blackwell theorem shows that any estimator can be improved by means of a minimum sufficient statistic.

Theorem 4 (*of Rao-Blackwell, see Cassel et al., 1993, p. 40*)**.** *Let $p(\mathbf{s})$ be a sampling design with replacement and an estimator $\widehat{T}$ (eventually biased) of T where $D^* = u^*(D)$. If $\widehat{T}^* = \mathrm{E}(\widehat{T}|D^*)$, then*

(i) $\mathrm{E}\left(\widehat{T}^*\right) = \mathrm{E}\left(\widehat{T}\right)$,

(ii) $\mathrm{MSE}\left(\widehat{T}\right) = \mathrm{MSE}\left(\widehat{T}^*\right) + \mathrm{E}\left(\widehat{T}^* - \widehat{T}\right)^2$,

(iii) $\mathrm{MSE}\left(\widehat{T}^*\right) \leq \mathrm{MSE}\left(\widehat{T}\right)$.

Proof. (i) is obvious because

$$\mathrm{E}(\widehat{T}) = \mathrm{EE}(\widehat{T}|D) = \mathrm{E}(\widehat{T}^*).$$

Next, for (ii), we have

$$\mathrm{MSE}\left(\widehat{T}\right) = \mathrm{E}\left(\widehat{T} - \widehat{T}^* + \widehat{T}^* - T\right)^2$$
$$= \mathrm{E}\left(\widehat{T} - \widehat{T}^*\right)^2 + \mathrm{E}\left(\widehat{T}^* - T\right)^2 - 2\mathrm{E}\left\{\left(\widehat{T} - \widehat{T}^*\right)\left(\widehat{T}^* - T\right)\right\}.$$

The third term is null because

$$\mathrm{E}\left\{\left(\widehat{T}^* - T\right)\left(\widehat{T} - \widehat{T}^*\right)\right\} = \mathrm{E}\left\{\left(\widehat{T}^* - T\right)\mathrm{E}\left(\widehat{T} - \widehat{T}^*|D\right)\right\} = 0.$$

Finally, (iii) comes directly from (ii). $\qquad\square$

The classical example of Rao-Blackwellization is the estimation of a total in simple random sampling with replacement and is developed in Section 4.6, page 53. The result about sufficiency is interesting but weak. The information about the multiplicity of the units can always be suppressed, but the sufficiency reduction does not permit the construction of an estimator.

Because it is impossible to identify an estimator by means of the technique of reduction by sufficiency, we are restricted to the use of linear unbiased estimators of the totals defined on page 21. Consider two cases.

- The first estimator takes into account the multiplicity of the units:

$$\widehat{Y}_{L1} = w_0 + \sum_{k \in U} w_k y_k S_k,$$

where $\mathrm{E}(w_0) = 0$ and $\mathrm{E}(w_k S_k) = 1$, for all $k \in U$. If the w_k's are not random, the only solution is: $w_0 = 0$ and $w_k = 1/\mathrm{E}(S_k) = 1/\mu_k$, for all $k \in U$, which gives the Hansen-Hurwitz estimator.

- The second estimator depends on the sample only through the reduction function:

$$\widehat{Y}_{L2} = w_0 + \sum_{k \in U} w_k y_k r(S_k),$$

where $\mathrm{E}(w_0) = 0$ and $\mathrm{E}[w_k r(S_k)] = 1$, for all $k \in U$. If the w_k's are not random, the only solution is: $w_0 = 0$ and $w_k = 1/\mathrm{E}[r(S_k)] = 1/\pi_k$, for all $k \in U$, which gives the Horvitz-Thompson estimator.

2.16 The Hansen-Hurwitz (HH) Estimator

2.16.1 Estimation of a Total

Definition 30. *The Hansen-Hurwitz estimator (see Hansen and Hurwitz, 1943) of Y is defined by*

$$\widehat{Y}_{HH} = \sum_{k \in U} \frac{S_k y_k}{\mu_k},$$

where $\mu_k = \mathrm{E}(S_k), k \in U$.

Result 7. *If $\mu_k > 0$, for all $k \in U$, then $\widehat{Y}_{HH}$ is an unbiased estimator of Y.*

Proof. If $\mu_k > 0$, for all $k \in U$, then

$$\mathrm{E}(\widehat{Y}_{HH}) = \sum_{k \in U} \frac{\mathrm{E}(S_k) y_k}{\mu_k} = \sum_{k \in U} y_k = Y. \qquad \square$$

2.16.2 Variance of the Hansen-Hurwitz Estimator

The variance of the Hansen-Hurwitz estimator is

$$\mathrm{var}_1(\widehat{Y}_{HH}) = \sum_{k \in U} \sum_{\ell \in U} \frac{y_k y_\ell \Sigma_{k\ell}}{\mu_k \mu_\ell}. \qquad (2.12)$$

Result 8. *When the design is of fixed sample size, the variance can also be written:*

$$\mathrm{var}_2(\widehat{Y}_{HH}) = -\frac{1}{2} \sum_{k \in U} \sum_{\substack{\ell \in U \\ \ell \neq k}} \left(\frac{y_k}{\mu_k} - \frac{y_\ell}{\mu_\ell} \right)^2 \Sigma_{k\ell}. \qquad (2.13)$$

Proof. We have

$$-\frac{1}{2} \sum_{k \in U} \sum_{\substack{\ell \in U \\ \ell \neq k}} \left(\frac{y_k}{\mu_k} - \frac{y_\ell}{\mu_\ell} \right)^2 \Sigma_{k\ell} = -\frac{1}{2} \sum_{k \in U} \sum_{\ell \in U} \left(\frac{y_k}{\mu_k} - \frac{y_\ell}{\mu_\ell} \right)^2 \Sigma_{k\ell}$$

$$= -\sum_{k \in U} \sum_{\ell \in U} \left(\frac{y_k^2}{\mu_k^2} \Sigma_{k\ell} - \frac{y_k y_\ell}{\mu_k \mu_\ell} \Sigma_{k\ell} \right) = \sum_{k \in U} \sum_{\ell \in U} \frac{y_k y_\ell}{\mu_k \mu_\ell} \Sigma_{k\ell} - \sum_{k \in U} \frac{y_k^2}{\mu_k^2} \sum_{\ell \in U} \Sigma_{k\ell}.$$

When the sampling design has a fixed sample size (see Result 3, page 16),

$$\sum_{\ell \in U} \Sigma_{k\ell} = 0,$$

which proves Result 8. $\qquad \square$

2.16.3 Variance Estimation

From expressions (2.12) and (2.13), it is possible to construct two variance estimators. For general sampling designs, we have:

$$\widehat{\mathrm{var}}_1(\widehat{Y}_{HH}) = \sum_{k \in U} \sum_{\ell \in U} \frac{S_k S_\ell y_k y_\ell \Sigma_{k\ell}}{\mu_k \mu_\ell \mu_{k\ell}}. \tag{2.14}$$

For sampling designs with fixed sample size, we have:

$$\widehat{\mathrm{var}}_2(\widehat{Y}_{HH}) = -\frac{1}{2} \sum_{k \in U} \sum_{\substack{\ell \in U \\ \ell \neq k}} \left(\frac{y_k}{\mu_k} - \frac{y_\ell}{\mu_\ell} \right)^2 \frac{S_k S_\ell \Sigma_{k\ell}}{\mu_{k\ell}}. \tag{2.15}$$

Both estimators are unbiased if $\mu_{k\ell} > 0$ for all $k, \ell \in U$. Indeed,

$$\mathrm{E}\left[\widehat{\mathrm{var}}_1(\widehat{Y}_{HH}) \right] = \sum_{k \in U} \sum_{\ell \in U} \frac{y_k y_\ell \Sigma_{k\ell}}{\mu_k \mu_\ell \mu_{k\ell}} \mathrm{E}(S_k S_\ell) = \mathrm{var}_1(\widehat{Y}_{HH}),$$

and

$$\mathrm{E}\left[\widehat{\mathrm{var}}_2(\widehat{Y}_{HH}) \right] = -\frac{1}{2} \sum_{k \in U} \sum_{\substack{\ell \in U \\ \ell \neq k}} \left(\frac{y_k}{\mu_k} - \frac{y_\ell}{\mu_\ell} \right)^2 \frac{\Sigma_{k\ell}}{\mu_{k\ell}} \mathrm{E}(S_k S_\ell) = \mathrm{var}_2(\widehat{Y}_{HH}).$$

By developing $\widehat{\mathrm{var}}_2(\widehat{Y}_{HH})$, we get the following result.

Result 9.

$$\widehat{\mathrm{var}}_1(\widehat{Y}_{HH}) = \widehat{\mathrm{var}}_2(\widehat{Y}_{HH}) + \sum_{k \in U} \frac{S_k y_k^2}{\mu_k^2} \sum_{\ell \in U} \frac{S_\ell \Sigma_{k\ell}}{\mu_{k\ell}}.$$

Note that, with fixed sample size, estimator (2.14) should never be used. Indeed, if $y_k = \mu_k, k \in U$ and the sampling design has a fixed sample size, we have

$$\mathrm{var}(\widehat{Y}_{HH}) = 0,$$

$$\widehat{\mathrm{var}}_2(\widehat{Y}_{HH}) = 0,$$

but

$$\widehat{\mathrm{var}}_1(\widehat{Y}_{HH}) = \sum_{k \in U} S_k \sum_{\ell \in U} \frac{S_\ell \Sigma_{k\ell}}{\mu_{k\ell}},$$

which is not generally equal to 0.

2.17 The Horvitz-Thompson (HT) Estimator

2.17.1 Estimation of a Total

The Horvitz-Thompson estimator (see Horvitz and Thompson, 1952) is defined by

$$\widehat{Y}_{HT} = \sum_{k \in U} \frac{r(S_k) y_k}{\pi_k} = r(\mathbf{S})' \check{\mathbf{y}},$$

where $\check{\mathbf{y}} = (y_1/\pi_1 \quad \cdots \quad y_k/\pi_k \quad \cdots \quad y_k/\pi_k)$. When sampling is without replacement, the Horvitz-Thompson estimator is equal to the Hansen-Hurwitz estimator.

2.17.2 Variance of the Horvitz-Thompson Estimator

The variance of the Horvitz-Thompson estimator is

$$\mathrm{var}_1(\widehat{Y}_{HT}) = \sum_{k \in U} \sum_{\ell \in U} \frac{y_k y_\ell \Delta_{k\ell}}{\pi_k \pi_\ell} = \check{\mathbf{y}}' \boldsymbol{\Delta} \check{\mathbf{y}}. \qquad (2.16)$$

When the size of the reduced sample $n[r(\mathbf{S})]$ is fixed, we can also write

$$\mathrm{var}_2(\widehat{Y}_{HT}) = -\frac{1}{2} \sum_{k \in U} \sum_{\substack{\ell \in U \\ \ell \neq k}} \left(\frac{y_k}{\pi_k} - \frac{y_\ell}{\pi_\ell} \right)^2 \Delta_{k\ell}. \qquad (2.17)$$

2.17.3 Variance Estimation

From expressions (2.16) and (2.17), it is possible to construct two variance estimators. For general sampling designs, we have:

$$\widehat{\mathrm{var}}_1(\widehat{Y}_{HT}) = \sum_{k \in U} \sum_{\ell \in U} \frac{r(S_k) r(S_\ell) y_k y_\ell \Delta_{k\ell}}{\pi_k \pi_\ell \pi_{k\ell}}. \qquad (2.18)$$

When the reduced sample size $n[r(\mathbf{S})]$ is fixed, we can define the Sen-Yates-Grundy estimator (see Sen, 1953; Yates and Grundy, 1953):

$$\widehat{\mathrm{var}}_2(\widehat{Y}_{HT}) = -\frac{1}{2} \sum_{k \in U} \sum_{\substack{\ell \in U \\ \ell \neq k}} \left(\frac{y_k}{\pi_k} - \frac{y_\ell}{\pi_\ell} \right)^2 \frac{r(S_k) r(S_\ell) \Delta_{k\ell}}{\pi_{k\ell}}. \qquad (2.19)$$

If $\pi_{k\ell} > 0$ for all $k, \ell \in U$, $\widehat{\mathrm{var}}_1(\widehat{Y}_{HT})$ and $\widehat{\mathrm{var}}_2(\widehat{Y}_{HT})$ are unbiased estimators of $\mathrm{var}_1(\widehat{Y}_{HT})$ and $\mathrm{var}_2(\widehat{Y}_{HT})$, respectively.

　　Both estimators can take negative values. Nevertheless, if $\Delta_{k\ell} < 0$, for all $k \neq \ell \in U$, then $\widehat{\mathrm{var}}_2(\widehat{Y}_{HT}) \geq 0$. Due to the double sum, both variance estimators are not really applicable. For each particular sampling design, either they can be simplified in a simple sum or an approximation must be used.

　　By developing $\widehat{\mathrm{var}}_2(\widehat{Y}_{HT})$, we get the following result.

Result 10.

$$\widehat{\text{var}}_1(\widehat{Y}_{HT}) = \widehat{\text{var}}_2(\widehat{Y}_{HT}) + \sum_{k \in U} \frac{r(S_k)y_k^2}{\pi_k^2} \sum_{\ell \in U} \frac{r(S_\ell)\Delta_{k\ell}}{\pi_{k\ell}}.$$

Note that, with fixed sample size, estimator (2.18) should never be used.

2.18 More on Estimation in Sampling With Replacement

When the sampling design is with replacement, there exist three basic ways to estimate the total:

1. The Hansen-Hurwitz estimator

$$\widehat{Y}_{HH} = \sum_{k \in U} \frac{S_k y_k}{\mu_k},$$

2. The Horvitz-Thompson estimator

$$\widehat{Y}_{HT} = \sum_{k \in U} \frac{r(S_k)y_k}{\pi_k},$$

3. The improved Hansen-Hurwitz estimator

$$\widehat{Y}_{IHH} = \text{E}\left[\widehat{Y}_{HH} | r(\mathbf{S})\right] = \sum_{k \in U} \frac{\text{E}[S_k | r(\mathbf{S})]y_k}{\mu_k}.$$

These three estimators are unbiased. They are equal when the sampling design is without replacement.

From Theorem 4 (page 25) of Rao-Blackwell, we have that

$$\text{var}(\widehat{Y}_{IHH}) \leq \text{var}(\widehat{Y}_{HH}). \tag{2.20}$$

The Hansen-Hurwitz estimator is not admissible, in the sense that it is always possible to construct (at least theoretically) an improved estimator that always has a smaller variance than $\widehat{Y}_{HH}$. Unfortunately, the conditional expectation $\text{E}[S_k | r(\mathbf{S})]$ needed to compute $\widehat{Y}_{IHH}$ is often very intricate, especially when the units are selected with unequal probabilities.

3

Sampling Algorithms

3.1 Introduction

A sampling algorithm is a procedure used to select a sample. If the sampling design is known and if the population size is not too large, a sample can be selected directly by enumerating all the samples as explained in Section 3.3. Nevertheless, when N is large, the number of possible samples becomes so large that it is practically impossible to enumerate all the samples. The objective of a sampling algorithm is to select a sample by avoiding the enumeration of all the samples. The main difficulty of the implementation is thus the combinatory explosion of the number of possible samples.

Hedayat and Sinha (1991) and Chen (1998) pointed out that several distinct sampling algorithms (or sampling schemes) can correspond to the same sampling design, which is clearly illustrated in Chapter 4, devoted to simple random sampling. There are, however, several families of algorithms to implement sampling designs. In this chapter, we describe these general algorithms and propose a typology. Indeed, most of the particular algorithms can be presented as particular cases of very general schemes that can be applied to any sampling design.

3.2 Sampling Algorithms

We refer to the following definitions:

Definition 31. *A sampling algorithm is a procedure allowing the selection of a random sample.*

Definition 32. *A sampling algorithm is said to be enumerative if all the possible samples must be listed in order to select the random sample.*

An efficient sampling algorithm is by definition a fast one. All enumerative algorithms are therefore inefficient. The basic aim is thus to find a shortcut that allows us to select a sample by means of a sampling design and to avoid the complete computation of the sample. It seems that there is no general method for constructing a shortcut for any sampling design. The possibility of constructing a fast algorithm depends on the sampling design that must be implemented. There exist, however, some standard algorithms of sampling that can be applied to any sampling design. Nevertheless, nonenumerative sampling designs are obtained only for very specific designs.

3.3 Enumerative Selection of the Sample

In order to implement a sampling design $p(.)$, an enumerative procedure presented in Algorithm 3.1 is theoretically always possible. As presented in Ta-

---

Algorithm 3.1 Enumerative algorithm

---

1. First, construct a list $\{s_1, s_2, \ldots, s_j, \ldots, s_J\}$ of all possible samples with their probabilities.
2. Next, generate a random variable u with a uniform distribution in $[0,1]$.
3. Finally, select the sample s_j such that $\sum_{i=1}^{j-1} p(s_i) \leq u < \sum_{i=1}^{j} p(s_i)$.

---

ble 3.1, even for small population sizes, the size of the support is so large that it becomes unworkable very quickly. A large part of the developments in sampling theory is the research of shortcuts for avoiding this enumeration.

Table 3.1. Sizes of symmetric supports

Support $\mathcal{Q}$	card($\mathcal{Q}$)	$N = 100, n = 10$	$N = 300, n = 30$
$\mathcal{R}$	∞	–	–
$\mathcal{R}_n$	$\binom{N+n-1}{n}$	5.1541×10^{13}	3.8254×10^{42}
$\mathcal{S}$	2^N	1.2677×10^{30}	2.0370×10^{90}
$\mathcal{S}_n$	$\binom{N}{n}$	1.7310×10^{13}	1.7319×10^{41}

3.4 Martingale Algorithms

A very large family is the set of the martingale algorithms. Almost all sampling methods can be expressed in the form of a martingale algorithm, except the rejective algorithm (on this topic see Section 3.8).

Definition 33. *A sampling algorithm is said to be a martingale algorithm if it can be written as a finite sequence of random vectors*

$$\boldsymbol{\mu}(0), \boldsymbol{\mu}(1), \boldsymbol{\mu}(2), \ldots, \boldsymbol{\mu}(t), \ldots, \boldsymbol{\mu}(T), \text{ of } \mathbb{R}^N,$$

such that

- $\boldsymbol{\mu}(0) = \boldsymbol{\mu} = \mathrm{E}(\mathbf{S}),$
- $\mathrm{E}\left[\boldsymbol{\mu}(t)|\boldsymbol{\mu}(t-1), \ldots, \boldsymbol{\mu}(0)\right] = \boldsymbol{\mu}(t-1),$
- $\boldsymbol{\mu}(t)$ *is in the convex hull of the support,*
- $\boldsymbol{\mu}(T) \in \mathcal{R}.$

A martingale algorithm is thus a "random walk" in the convex hull of the support that begins at the inclusion probability vector and stops on a sample.

Definition 34. *In sampling without replacement, a sampling algorithm is said to be a martingale algorithm if it can be written as a finite sequence of random vectors*

$$\boldsymbol{\pi}(0), \boldsymbol{\pi}(1), \boldsymbol{\pi}(2), \ldots, \boldsymbol{\pi}(t), \ldots, \boldsymbol{\pi}(T), \text{ of } [0,1]^N,$$

such that

- $\boldsymbol{\pi}(0) = \boldsymbol{\pi} = \mathrm{E}(\mathbf{S}),$
- $\mathrm{E}\left[\boldsymbol{\pi}(t)|\boldsymbol{\pi}(t-1), \ldots, \boldsymbol{\pi}(0)\right] = \boldsymbol{\pi}(t-1),$
- $\boldsymbol{\pi}(t)$ *is in the convex hull of the support,*
- $\boldsymbol{\pi}(T) \in \mathcal{S}.$

Note that, in sampling without replacement, the convex hull of the support is a subset of the hypercube $[0,1]^N$.

A martingale algorithm consists thus of modifying randomly at each step the vector of inclusion probabilities until a sample is obtained. All the sequential procedures (see Algorithm 3.2, page 35), the draw by draw algorithms (see Algorithm 3.3, page 35), the algorithms of the splitting family (see Chapter 6), and the cube family of algorithms (see Chapter 8) can be presented as martingales.

3.5 Sequential Algorithms

A sequential procedure is a method that is applied to a list of units sorted according to a particular order denoted $1, \ldots, k, \ldots, N$.

Definition 35. *A sampling procedure is said to be weakly sequential if at step $k = 1, \ldots, N$ of the procedure, the decision concerning the number of times that unit k is in the sample is definitively taken.*

Definition 36. *A sampling procedure is said to be strictly sequential if it is weakly sequential and if the decision concerning unit k does not depend on the units that are after k on the list.*

The standard sequential procedure consists of examining successively all the units of the population. At each one of the N steps, unit k is selected s_k times according to a distribution probability that depends on the decision taken for the previous units. The standard sequential procedure presented in Algorithm 3.2 is a weakly sequential method that can be defined and can theoretically be used to implement any sampling design.

Algorithm 3.2 Standard sequential procedure

---

1. Let $p(\mathbf{s})$ be the sampling design and $\mathcal{Q}$ the support. First, define

$$q_1(s_1) = \Pr(S_1 = s_1) = \sum_{\mathbf{s} \in \mathcal{Q} | S_1 = s_1} p(\mathbf{s}), s_1 = 0, 1, 2, \ldots$$

2. Select the first unit s_1 times according to the distribution $q_1(s_1)$.
3. For $k = 2, \ldots, N$ do
   a) Compute

$$q_k(s_k) = \Pr(S_k = s_k | S_{k-1} = s_{k-1}, \ldots, S_1 = s_1)$$
$$= \frac{\sum_{\mathbf{s} \in \mathcal{Q} | S_k = s_k, S_{k-1} = s_{k-1}, \ldots, S_1 = s_1} p(\mathbf{s})}{\sum_{\mathbf{s} \in \mathcal{Q} | S_{k-1} = s_{k-1}, \ldots, S_1 = s_1} p(\mathbf{s})}, s_k =, 0, 1, 2, \ldots$$

   b) Select the kth unit s_k times according to the distribution $q_k(s_k)$;
   EndFor.

This procedure provides a nonenumerative procedure only if the $q_k(s_k)$ can be computed successively without enumerating all the possible samples of the sampling designs. The standard sequential algorithm is implemented for particular sampling designs in Algorithm 4.1, page 44; Algorithm 4.3, page 48; Algorithm 4.6, page 53; Algorithm 4.8, page 61; Algorithm 5.1, page 70; Algorithm 5.2, page 75; Algorithm 5.4, page 79; and Algorithm 5.8, page 92.

Example 4. Suppose that the sampling design is

$$p(\mathbf{s}) = \binom{N}{n}^{-1}, \mathbf{s} \in \mathcal{S}_n,$$

which is actually a simple random sampling without replacement (see Section 4.4). After some algebra, we get

$$q_k(s_k) = \begin{cases} \dfrac{n - \sum_{\ell=1}^{k-1} s_\ell}{N - k + 1} & \text{if } s_k = 1 \\ 1 - \dfrac{n - \sum_{\ell=1}^{k-1} s_\ell}{N - k + 1} & \text{if } s_k = 0. \end{cases}$$

This method was actually proposed by Fan et al. (1962) and Bebbington (1975) and is described in detail in Algorithm 4.3, page 48.

3.6 Draw by Draw Algorithms

The draw by draw algorithms are restricted to designs with fixed sample size. We refer to the following definition.

Definition 37. *A sampling design of fixed sample size n is said to be draw by draw if, at each one of the n steps of the procedure, a unit is definitively selected in the sample.*

For any sampling design $p(\mathbf{s})$ of fixed sample size on a support $\mathcal{Q} \subset \mathcal{R}_n$, there exists a standard way to implement a draw by draw procedure. At each one of the n steps, a unit is selected randomly from the population with probabilities proportional to the mean vector. Next, a new sampling design is computed according to the selected unit. This procedure can be applied to sampling with and without replacement and is presented in Algorithm 3.3.

Algorithm 3.3 Standard draw by draw algorithm

---

1. Let $p(\mathbf{s})$ be a sampling design and $\mathcal{Q} \subset \mathcal{R}_n$ the support. First, define $p^{(0)}(\mathbf{s}) = p(\mathbf{s})$ and $\mathcal{Q}(0) = \mathcal{Q}$. Define also $\mathbf{b}(0)$ as the null vector of $\mathbb{R}^N$.
2. FOR $t = 0, \ldots, n-1$ DO
   a) Compute $\boldsymbol{\nu}(t) = \sum_{\mathbf{s} \in \mathcal{Q}(t)} \mathbf{s} p^{(t)}(\mathbf{s})$;
   b) Select randomly one unit from U with probabilities $q_k(t)$, where

   $$q_k(t) = \frac{\nu_k(t)}{\sum_{\ell \in U} \nu_\ell(t)} = \frac{\nu_k(t)}{n-t}, k \in U;$$

   The selected unit is denoted j;
   c) Define $\mathbf{a}_j = (0 \cdots 0 \underbrace{1}_{j\text{th}} 0 \cdots 0)$; Execute $\mathbf{b}(t+1) = \mathbf{b}(t) + \mathbf{a}_j$;
   d) Define $\mathcal{Q}(t+1) = \{\tilde{\mathbf{s}} = \mathbf{s} - \mathbf{a}_j, \text{ for all } \mathbf{s} \in \mathcal{Q}(t) \text{ such that } s_j > 0\}$;
   e) Define, for all $\tilde{\mathbf{s}} \in \mathcal{Q}(t+1)$,

   $$p^{(t+1)}(\tilde{\mathbf{s}}) = \frac{s_j p^{(t)}(\mathbf{s})}{\sum_{\mathbf{s} \in \mathcal{Q}(t)} s_j p^{(t)}(\mathbf{s})},$$

   where $\mathbf{s} = \tilde{\mathbf{s}} + \mathbf{a}_j$;
   ENDFOR.
3. The selected sample is $\mathbf{b}(n)$.

The validity of the method is based on the recursive relation:

$$p^{(t)}(\mathbf{s}) = \sum_{k \in U} q_k(t) \frac{s_j p^{(t)}(\mathbf{s})}{\sum_{\mathbf{s} \in \mathcal{Q}(t)} s_j p^{(t)}(\mathbf{s})} = \sum_{k \in U} q_k(t) p^{(t+1)}(\tilde{\mathbf{s}}),$$

where $\mathbf{s} \subset \mathcal{Q}(t)$, and $\tilde{\mathbf{s}} = \mathbf{s} - \mathbf{a}_j$. As $\mathcal{Q}(0) \subset \mathcal{R}_n$, we have $\mathcal{Q}(t) \subset \mathcal{R}_{n-t}$. Thus,

$$\sum_{k \in U} \nu_k(t) = (n - t).$$

Moreover, the sequence

$$\boldsymbol{\mu}(t) = \boldsymbol{\nu}(t) + \sum_{j=1}^{t} \mathbf{b}(j),$$

for $t = 0, \ldots, n - 1$, is a martingale algorithm.

The standard draw by draw procedure can theoretically be implemented for any sampling design but is not necessarily nonenumerative. In order to provide a nonenumerative algorithm, the passage from $q_k(t)$ to $q_k(t+1)$ should be such that it is not necessary to compute $p^{(t)}(\mathbf{s})$, which depends on the sampling design and on the support. The standard draw by draw algorithm is implemented for particular sampling designs in Algorithm 4.2, page 48; Algorithm 4.7, page 60; Algorithm 5.3, page 76; and Algorithm 5.9, page 95.

Example 5. Suppose that the sampling design is

$$p(\mathbf{s}) = \frac{n!}{N^n} \prod_{k \in U} \frac{1}{s_k!}, \quad \mathbf{s} \in \mathcal{R}_n,$$

which is actually a simple random sampling with replacement (see Section 4.6). We have, for all $t = 0, \ldots, n - 1$,

$$\boldsymbol{\nu}(t) = \left(\frac{n - t}{N} \quad \cdots \quad \frac{n - t}{N} \right)',$$

$$q_k(t) = \frac{1}{N}, k \in U,$$

$$\mathcal{Q}(t) = \mathcal{R}_{n-t},$$

$$p^{(t)}(\mathbf{s}) = \frac{(n - t)!}{N^{(n-t)}} \prod_{k \in U} \frac{1}{s_k!}, \mathbf{s} \in \mathcal{R}_{n-t}.$$

It is not even necessary to compute $p^{(t)}(\mathbf{s})$. At each step, a unit is selected from U with probability $1/N$ for each unit. This method is developed in Algorithm 4.7, page 60.

In the case of sampling without replacement, the standard draw by draw algorithm can be simplified as presented in Algorithm 3.4.

Example 6. Suppose that the sampling design is

$$p(\mathbf{s}) = \binom{N}{n}^{-1}, \quad \mathbf{s} \in \mathcal{S}_n,$$

Algorithm 3.4 Standard draw by draw algorithm for sampling without replacement

1. Let $p(\mathbf{s})$ be a sampling design and $\mathcal{Q} \in \mathcal{S}$ the support.
2. Define $\mathbf{b} = (b_k) = \mathbf{0} \in \mathbb{R}^N$.
3. FOR $t = 0, \ldots, n-1$ DO
    select a unit from U with probability

$$q_k = \begin{cases} \dfrac{1}{n-t} \mathrm{E}\left(S_k | S_i = 1 \text{ for all } i \text{ such that } b_i = 1\right) & \text{if } b_k = 0 \\ 0 & \text{if } b_k = 1; \end{cases}$$

    IF unit j is selected, THEN $b_j = 1$;
    ENDFOR.

which is actually a simple random sampling without replacement (see Section 4.4). In sampling without replacement, at each step, a unit is definitively selected in the sample. This unit cannot be selected twice. The draw by draw standard method is:

$$\nu_k(t) = \begin{cases} \dfrac{n-t}{N-t} & \text{if unit } k \text{ is not yet selected} \\ 0 & \text{if unit } k \text{ is already selected,} \end{cases}$$

$$q_k(t) = \begin{cases} \dfrac{1}{N-t} & \text{if unit } k \text{ is not yet selected} \\ 0 & \text{if unit } k \text{ is already selected.} \end{cases}$$

Thus, at step t, a unit is selected with equal probabilities $1/(N-t)$ among the $N-t$ unselected units. The vector $\boldsymbol{\pi}(t) = [\pi_k(t)] = $ is a martingale, where

$$\pi_k(t) = \begin{cases} \dfrac{1}{N-t} & \text{if unit } k \text{ is not yet selected} \\ 1 & \text{if unit } k \text{ is already selected,} \end{cases}$$

for $t = 0, \ldots, n-1$. This method is developed in Algorithm 4.2, page 48.

3.7 Eliminatory Algorithms

For sampling without replacement and of fixed sample size, it is possible to use methods for which units are eliminated from the population, which can be defined as follows.

Definition 38. *A sampling algorithm of fixed sample size n is said to be eliminatory if, at each one of the $N-n$ steps of the algorithm, a unit is definitively eliminated from the population.*

Algorithm 3.5 Standard eliminatory algorithm

---

1. Let $p(\mathbf{s})$ be the sampling design and $\mathcal{Q}$ the support such that $\mathcal{Q} \subset \mathcal{S}_n$. Define $p^{(0)}(\mathbf{s}) = p(\mathbf{s})$ and $\mathcal{Q}(0) = \mathcal{Q}$.

2. FOR $t = 0, \ldots, N - n - 1$ DO

   a) Compute $\boldsymbol{\nu}(t) = [\nu_k(t)]$, where

$$\nu_k(t) = \sum_{\mathbf{s} \in \mathcal{Q}(t) | s_k = 0} p^{(t)}(\mathbf{s}) = \sum_{\mathbf{s} \in \mathcal{Q}(t)} (1 - s_k) p^{(t)}(\mathbf{s});$$

   b) Select randomly a unit from U with unequal probabilities $q_k(t)$, where

$$q_k(t) = \frac{\nu_k(t)}{\sum_{\ell \in U} \nu_\ell(t)} = \frac{\nu_k(t)}{N - n - t}, \quad k \in U;$$

   The selected unit is denoted j;

   c) Define $\mathcal{Q}(t + 1) = \{\mathbf{s} \in \mathcal{Q}(t) | s_j = 0\}$;

   d) Define

$$p^{(t+1)}(\mathbf{s}) = p^{(t)}(\mathbf{s} | \mathcal{Q}(t + 1)) = \frac{p^{(t)}(\mathbf{s})}{\sum_{\mathbf{s} \in \mathcal{Q}(t+1)} p^{(t)}(\mathbf{s})},$$

   where $\mathbf{s} \in \mathcal{Q}(t + 1)$;

   ENDFOR.

3. The support $\mathcal{Q}(N - n)$ contains only one sample that is selected.

Algorithm 3.5 implements a standard eliminatory procedure for a fixed size sampling design without replacement.

The eliminatory methods can be presented as the complementary procedures of the draw by draw methods. Indeed, in order to implement an eliminatory method for a sampling design $p(.)$, we can first define the complementary design $p^c(.)$ of $p(.)$. Next, a sample $\mathbf{s}$ of size $N - n$ is randomly selected by means of a draw by draw method for $p^c(.)$. Finally, the sample $\mathbf{1} - \mathbf{s}$ is selected, where $\mathbf{1}$ is a vector of N ones.

3.8 Rejective Algorithms

Definition 39. *Let $\mathcal{Q}_1$ and $\mathcal{Q}_2$ be two supports such that $\mathcal{Q}_1 \subset \mathcal{Q}_2$. Also, let $p_1(.)$ be a sampling design on $\mathcal{Q}_1$ and $p_2(.)$ be a sampling design on $\mathcal{Q}_2$, such that*

$$p_2(\mathbf{s} | \mathcal{Q}_1) = p_1(\mathbf{s}), \quad \text{for all } \mathbf{s} \in \mathcal{Q}_1.$$

A sampling design is said to be rejective if successive independent samples are selected randomly according to $p_2(.)$ in $\mathcal{Q}_2$ until a sample of $\mathcal{Q}_1$ is obtained.

Example 7. Suppose that we want to select a sample with the design

$$p_1(\mathbf{s}) = \binom{N}{n}^{-1}, \quad \mathbf{s} \in \mathcal{S}_n,$$

which is a simple random sampling without replacement. Independent samples can be selected with the sampling design

$$p_2(\mathbf{s}) = \frac{n!}{N^n} \prod_{k \in U} \frac{1}{s_k!}, \quad \mathbf{s} \in \mathcal{R}_n,$$

which is a simple random sampling with replacement, until a sample of $\mathcal{S}_n$ is obtained. Indeed, we have

$$p_2(\mathbf{s}|\mathcal{S}_n) = p_1(\mathbf{s}), \quad \text{for all } \mathbf{s} \in \mathcal{S}_n.$$

In practice, rejective algorithms are quite commonly used. The rejective methods are sometimes very fast, and sometimes very slow according to the cardinality difference between $\mathcal{Q}_1$ and $\mathcal{Q}_2$. Rejective algorithms are implemented for particular sampling designs in Algorithm 5.5, page 89; Algorithm 5.6, page 90; Algorithm 7.3, page 136; Algorithm 7.4, page 136; and Algorithm 7.5, page 136.

4

Simple Random Sampling

4.1 Introduction

In this chapter, a unified theory of simple random sampling is presented. First, an original definition of a simple design is proposed. All the particular simple designs as simple random sampling with and without replacement and Bernoulli sampling with and without replacement can then be deduced by a simple change of symmetric support.

For each of these sampling designs, the standard algorithms presented in Chapter 3 are applied, which allows defining eight sampling procedures. It is interesting to note that these algorithms were originally published in statistics and computer science journals without many links between the publications of these fields of research.

In sampling with replacement with fixed sample size, the question of estimation is largely developed. It is indeed preferable to suppress the information about the multiplicity of the units, which amounts to applying a Rao-Blackwellization on the Hansen-Hurwitz estimator. The interest of each estimator is thus discussed.

4.2 Definition of Simple Random Sampling

Curiously, a concept as common as simple random sampling is often not defined. We refer to the following definition.

Definition 40. *A sampling design* $p_{\mathrm{SIMPLE}}(., \theta, \mathcal{Q})$ *of parameter* $\theta \in \mathbb{R}_+^*$ *on a support* $\mathcal{Q}$ *is said to be simple, if*

(i) Its sampling design can be written

$$p_{\mathrm{SIMPLE}}(\mathbf{s}, \theta, \mathcal{Q}) = \frac{\theta^{n(\mathbf{s})} \prod_{k \in U} 1/s_k!}{\sum_{\mathbf{s} \in \mathcal{Q}} \theta^{n(\mathbf{s})} \prod_{k \in U} 1/s_k!}, \quad \text{for all } \mathbf{s} \in \mathcal{Q}.$$

(ii) Its support $\mathcal{Q}$ *is symmetric (see Definition 2, page 9).*

Remark 7. If a simple sampling design is defined on a support $\mathcal{Q}$ without replacement, then it can be written:

$$p_{\text{SIMPLE}}(\mathbf{s}, \theta, \mathcal{Q}) = \frac{\theta^{n(\mathbf{s})}}{\sum_{\mathbf{s} \in \mathcal{Q}} \theta^{n(\mathbf{s})}},$$

for all $\mathbf{s} \in \mathcal{Q}$.

Remark 8. If a simple sampling design has a fixed sample size, then it does not depend on the parameter anymore:

$$p_{\text{SIMPLE}}(\mathbf{s}, \theta, \mathcal{Q}) = \frac{\prod_{k \in U} 1/s_k!}{\sum_{\mathbf{s} \in \mathcal{Q}} \prod_{k \in U} 1/s_k!},$$

for all $\mathbf{s} \in \mathcal{Q}$.

Remark 9. If a simple sampling design is without replacement and of fixed sample size, then:

$$p_{\text{SIMPLE}}(\mathbf{s}, \theta, \mathcal{Q}) = \frac{1}{\operatorname{card}\mathcal{Q}},$$

for all $\mathbf{s} \in \mathcal{Q}$.

Remark 10. For any $\mathbf{s}^*$ obtained by permuting the elements of $\mathbf{s}$, we have

$$p_{\text{SIMPLE}}(\mathbf{s}|\mathcal{Q}) = p_{\text{SIMPLE}}(\mathbf{s}^*|\mathcal{Q}).$$

Remark 11. If $\mathbf{S}$ is a random sample selected by means of a simple random sampling on $\mathcal{Q}$, then, because the support is symmetric,

1. $\mu = \mathrm{E}(S_k)$ and $\pi = \mathrm{E}[r(S_k)]$ do not depend on k,
2. $\mu = \dfrac{\mathrm{E}[n(\mathbf{S})]}{N}$,
3. $\pi = \dfrac{\mathrm{E}\{n[r(\mathbf{S})]\}}{N}$.

The most studied simple sampling designs are:

- Bernoulli sampling design (BERN) on $\mathcal{S}$,
- Simple random sampling without replacement (with fixed sample size) (SRSWOR) on $\mathcal{S}_n$,
- Bernoulli sampling with replacement (BERNWR) on $\mathcal{R}$,
- Simple random sampling with replacement with fixed sample size (SRSWR) on $\mathcal{R}_n$.

4.3 Bernoulli Sampling (BERN)

4.3.1 Sampling Design

Definition 41. *A simple design defined with parameter θ on support $\mathcal{S} = \{0,1\}^N$ is called a Bernoulli sampling design with inclusion probabilities $\pi = \theta/(1+\theta)$.*

The Bernoulli sampling design can be deduced from Definition 40, page 41. Indeed,

$$
p_{\mathrm{BERN}}\left(\mathbf{s}, \pi = \frac{\theta}{1+\theta}\right)
$$

$$
= p_{\mathrm{SIMPLE}}(\mathbf{s}, \theta, \mathcal{S}) = \frac{\theta^{n(\mathbf{s})}}{\sum_{\mathbf{s}\in\mathcal{S}}\theta^{n(\mathbf{s})}} = \frac{\theta^{n(\mathbf{s})}}{\sum_{r=0}^{N}\sum_{\mathbf{s}\in\mathcal{S}_r}\theta^r} = \frac{\theta^{n(\mathbf{s})}}{\sum_{r=0}^{N}\binom{N}{r}\theta^r}
$$

$$
= \frac{\theta^{n(\mathbf{s})}}{(\theta+1)^N} = \left(\frac{\theta}{1+\theta}\right)^{n(\mathbf{s})}\left(1-\frac{\theta}{1+\theta}\right)^{N-n(\mathbf{s})} = \pi^{n(\mathbf{s})}(1-\pi)^{N-n(\mathbf{s})},
$$

for all $\mathbf{s} \in \mathcal{S}$, where $\pi = \theta/(1+\theta)$, $\pi \in]0,1[$. Bernoulli sampling design can also be written

$$
p_{\mathrm{BERN}}(\mathbf{s}, \pi) = \prod_{k\in U} \pi^{s_k}(1-\pi)^{1-s_k},
$$

where the $S_1,\ldots,S_N$ are independent and identically distributed Bernoulli variables with parameter π.

The characteristic function is

$$
\phi_{\mathrm{BERN}}(\mathbf{t})
$$

$$
= \sum_{\mathbf{s}\in\mathcal{S}} e^{i\mathbf{t}'\mathbf{s}}p_{\mathrm{BERN}}(\mathbf{s},\pi) = \sum_{\mathbf{s}\in\mathcal{S}} e^{i\mathbf{t}'\mathbf{s}}\pi^{n(\mathbf{s})}(1-\pi)^{N-n(\mathbf{s})}
$$

$$
= (1-\pi)^N \sum_{\mathbf{s}\in\mathcal{S}}\prod_{k\in U}\left(\frac{\pi}{1-\pi}e^{it_k}\right)^{s_k} = (1-\pi)^N \prod_{k\in U}\left(1+\frac{\pi}{1-\pi}e^{it_k}\right)
$$

$$
= \prod_{k\in U}\left(1-\pi+\pi e^{it_k}\right).
$$

By differentiating the characteristic function, we get the inclusion probabilities $\mathrm{E}(S_k) = \pi$, for all $k \in U$. By differentiating the characteristic function twice, we get the joint inclusion probabilities $\mathrm{E}(S_k S_\ell) = \pi^2$, for all $k \neq \ell \in U$.

The sample size distribution is binomial, $n(\mathbf{S}) \sim \mathcal{B}(N,\pi)$, and

$$
\Pr[n(\mathbf{s}) = r] = \sum_{\mathbf{s}\in\mathcal{S}_r} \pi^r(1-\pi)^{N-r} = \binom{N}{r}\pi^r(1-\pi)^{N-r},
$$

with $r = 0, \ldots, N$.

Because the S_k's have a Bernoulli distribution, the variance is

$$\mathrm{var}(S_k) = \pi(1 - \pi), \text{ for all } k \in U.$$

Due to the independence of the S_k,

$$\Delta_{k\ell} = \mathrm{cov}(S_k, S_\ell) = 0, \text{ for all } k \neq \ell \in U.$$

The variance-covariance operator is $\boldsymbol{\Sigma} = \pi(1 - \pi)\mathbf{I}$, where $\mathbf{I}$ is an $(N \times N)$ identity matrix.

4.3.2 Estimation

The Horvitz-Thompson estimator is

$$\widehat{Y}_{HT} = \frac{1}{\pi} \sum_{k \in U} y_k S_k.$$

The variance of the Horvitz-Thompson estimator is

$$\mathrm{var}(\widehat{Y}_{HT}) = \mathbf{\check{y}}' \boldsymbol{\Delta} \mathbf{\check{y}} = \frac{\pi(1 - \pi)}{\pi^2} \sum_{k \in U} y_k^2.$$

The variance estimator of the Horvitz-Thompson estimator is

$$\widehat{\mathrm{var}}(\widehat{Y}_{HT}) = \frac{1 - \pi}{\pi^2} \sum_{k \in U} y_k^2 S_k.$$

4.3.3 Sequential Sampling Procedure for BERN

Due to the independence of the sampling between the units, the standard sequential Algorithm 3.2, page 34, is very easy to construct for a BERN sampling as presented in Algorithm 4.1.

Algorithm 4.1 Bernoulli sampling without replacement

DEFINITION k : INTEGER;
FOR $k = 1, \ldots, N$ DO with probability π select unit k; ENDFOR.

Note that Algorithm 4.1 is strictly sequential. Conditionally on a fixed sample size, a Bernoulli design becomes a simple random sample without replacement. Another method for selecting a sample with a Bernoulli design is thus to choose randomly the sample size according to a binomial distribution and then to select a sample with a simple random sampling without replacement.

4.4 Simple Random Sampling Without Replacement (SRSWOR)

4.4.1 Sampling Design

Definition 42. *A simple design defined with parameter θ on support $\mathcal{S}_n$ is called a simple random sampling without replacement.*

The simple random sampling without replacement can be deduced from Definition 40, page 41:

$$
p_{\text{SRSWOR}}(\mathbf{s}, n) = p_{\text{SIMPLE}}(\mathbf{s}, \theta, \mathcal{S}_n) = \frac{\theta^n}{\sum_{\mathbf{s} \in \mathcal{S}_n} \theta^n}
$$

$$
= \frac{1}{\text{card}(\mathcal{S}_n)} = \binom{N}{n}^{-1} = \frac{n!(N-n)!}{N!},
$$

for all $\mathbf{s} \in \mathcal{S}_n$. Note that $p_{\text{SRSWOR}}(\mathbf{s}, n)$ does not depend on θ anymore. The characteristic function is

$$
\phi_{\text{SRSWOR}}(\mathbf{t}) = \sum_{\mathbf{s} \in \mathcal{S}_n} e^{i \mathbf{t}' \mathbf{s}} p_{\text{SRSWOR},n}(\mathbf{s}) = \binom{N}{n}^{-1} \sum_{\mathbf{s} \in \mathcal{S}_n} e^{i \mathbf{t}' \mathbf{s}},
$$

and cannot be simplified. The inclusion probability is

$$
\pi_k = \sum_{\mathbf{s} \in \mathcal{S}_n} s_k p_{\text{SRSWOR}}(\mathbf{s}, n) = \text{card}\{\mathbf{s} \in \mathcal{S}_n | s_k = 1\} \binom{N}{n}^{-1}
$$

$$
= \binom{N-1}{n-1} \binom{N}{n}^{-1} = \frac{n}{N},
$$

for all $k \in U$.

The joint inclusion probability is

$$
\pi_{k\ell} = \sum_{\mathbf{s} \in \mathcal{S}_n} s_k s_\ell p_{\text{SRSWOR}}(\mathbf{s}, n) = \text{card}\{\mathbf{s} \in \mathcal{S}_n | s_k = 1 \text{ and } s_\ell = 1\} \binom{N}{n}^{-1}
$$

$$
= \binom{N-2}{n-2} \binom{N}{n}^{-1} = \frac{n(n-1)}{N(N-1)},
$$

for all $k \neq \ell \in U$.

The variance-covariance operator is given by

$$
\Delta_{k\ell} = \text{cov}(S_k, S_\ell) = \begin{cases} \pi_k(1 - \pi_k) = \dfrac{n(N-n)}{N^2}, & k = \ell \in U \\[2mm] \dfrac{n(n-1)}{N(N-1)} - \dfrac{n^2}{N^2} = -\dfrac{n(N-n)}{N^2(N-1)}, & k \neq \ell \in U. \end{cases}
$$

Thus,

$$\mathbf{\Delta} = \mathrm{var}(\mathbf{S}) = \frac{n(N-n)}{N(N-1)}\mathbf{H},$$

where $\mathbf{H}$ is the projection matrix that centers the data.

$$\mathbf{H} = \mathbf{I} - \frac{\mathbf{1}\mathbf{1}'}{N}, = \begin{pmatrix} 1 - \frac{1}{N} & \cdots & \frac{-1}{N} & \cdots & \frac{-1}{N} \\ \vdots & \ddots & \vdots & & \vdots \\ \frac{-1}{N} & \cdots & 1-\frac{1}{N} & \cdots & \frac{-1}{N} \\ \vdots & & \vdots & \ddots & \vdots \\ \frac{-1}{N} & \cdots & \frac{-1}{N} & \cdots & 1-\frac{1}{N} \end{pmatrix}, \tag{4.1}$$

and $\mathbf{I}$ is an identity matrix and $\mathbf{1}$ is a vector of N ones. We have

$$\mathbf{H}\check{\mathbf{y}} = \frac{N}{n}\left(y_1 - \overline{Y} \quad \cdots \quad y_k - \overline{Y} \quad \cdots \quad y_N - \overline{Y}\right)',$$

where $\check{\mathbf{y}} = \frac{N}{n}\left(y_1 \quad \cdots \quad y_k \quad \cdots \quad y_N\right)'$. Thus,

$$\check{\mathbf{y}}'\mathbf{H}\check{\mathbf{y}} = \frac{N^2}{n^2}\sum_{k \in U}\left(y_k - \overline{Y}\right)^2.$$

In order to estimate the variance, we need to compute

$$\frac{\Delta_{k\ell}}{\pi_{k\ell}} = \frac{n(N-n)}{N(n-1)}\begin{cases} \dfrac{n-1}{n}, & k = \ell \in U \\ -\dfrac{1}{n}, & k \neq \ell \in U. \end{cases}$$

4.4.2 Estimation

Horvitz-Thompson estimator

The Horvitz-Thompson estimator is

$$\widehat{Y}_{HT} = \frac{N}{n}\sum_{k \in U} y_k S_k.$$

Variance of the Horvitz-Thompson estimator

The variance of the Horvitz-Thompson estimator is

$$\mathrm{var}(\widehat{Y}_{HT}) = \check{\mathbf{y}}'\mathbf{\Delta}\check{\mathbf{y}} = \frac{n(N-n)}{N(N-1)}\check{\mathbf{y}}'\mathbf{H}\check{\mathbf{y}} = \frac{n(N-n)}{N(N-1)}\frac{N^2}{n^2}\sum_{k \in U}\left(y_k - \overline{Y}\right)^2$$

$$= N^2\frac{(N-n)}{N}\frac{1}{n(N-1)}\sum_{k \in U}\left(y_k - \overline{Y}\right)^2 = N^2\frac{N-n}{N}\frac{V_y^2}{n},$$

where

$$V_y^2 = \frac{1}{N-1}\sum_{k \in U}\left(y_k - \overline{Y}\right)^2.$$

Variance estimator of the Horvitz-Thompson estimator

The variance estimator of the Horvitz-Thompson estimator is

$$\widehat{\mathrm{var}}(\widehat{Y}_{HT}) = N^2 \frac{N-n}{nN} v_y^2, \tag{4.2}$$

where

$$v_y^2 = \frac{1}{n-1} \sum_{k \in U} S_k \left(y_k - \frac{\widehat{Y}_{HT}}{N} \right)^2.$$

Note that (4.2) is a particular case of estimators (2.18) and (2.19), page 28, that are equal in this case. Estimator (4.2) is thus unbiased, but a very simple proof can be given for this particular case:

Result 11. *In a SRSWOR v_y^2 is an unbiased estimator of V_y^2.*

Proof. Because

$$v_y^2 = \frac{1}{n-1} \sum_{k \in U} S_k \left(y_k - \frac{\widehat{Y}_{HT}}{N} \right)^2 = \frac{1}{2n(n-1)} \sum_{k \in U} \sum_{\ell \in U} (y_k - y_\ell)^2 S_k S_\ell,$$

we obtain

$$\mathrm{E}(v_y^2) = \frac{1}{2n(n-1)} \sum_{k \in U} \sum_{\ell \in U} (y_k - y_\ell)^2 \, \mathrm{E}\left(S_k S_\ell\right)$$

$$= \frac{1}{2n(n-1)} \sum_{k \in U} \sum_{\ell \in U} (y_k - y_\ell)^2 \, \pi_{k\ell}$$

$$= \frac{1}{2n(n-1)} \sum_{k \in U} \sum_{\ell \in U} (y_k - y_\ell)^2 \, \frac{n(n-1)}{N(N-1)}$$

$$= \frac{1}{2N(N-1)} \sum_{k \in U} \sum_{\ell \in U} (y_k - y_\ell)^2 = V_y^2. \qquad \square$$

4.4.3 Draw by Draw Procedure for SRSWOR

The standard draw by draw Algorithm 3.3, page 35, becomes in the case of SRSWOR the well-known ball-in-urn method (see Example 4, page 34) where a unit is selected from the population with equal probability. The selected unit is removed from the population. Next, a second unit is selected among the $N-1$ remaining units. This operation is repeated n times and, when a unit is selected, it is definitively removed from the population. This method is summarized in Algorithm 4.2. Nevertheless, Algorithm 4.2 has an important drawback: its implementation is quite complex and it is not sequential. For this reason, it is preferable to use the selection-rejection procedure presented in Section 4.4.4.

Algorithm 4.2 Draw by draw procedure for SRSWOR

DEFINITION j : INTEGER;
FOR $t = 0, \ldots, n-1$ DO
    select a unit k from the population with probability
$$q_k = \begin{cases} \frac{1}{N-t} & \text{if } k \text{ is not already selected} \\ 0 & \text{if } k \text{ is already selected;} \end{cases}$$
ENDFOR.

4.4.4 Sequential Procedure for SRSWOR: The Selection-Rejection Method

The selection-rejection procedure is the standard sequential Algorithm 3.2, page 34, for SRSWOR; that is, the sample can be selected in one pass in the data file. The method was discussed in Fan et al. (1962), Bebbington (1975), Vitter (1984), Ahrens and Dieter (1985) and Vitter (1987) and is presented in Algorithm 4.3. The selection-rejection method is the best algorithm for selecting a sample according to SRSWOR; it is strictly sequential. Note that the population size must be known before applying Algorithm 4.3.

Algorithm 4.3 Selection-rejection procedure for SRSWOR

DEFINITION k, j : INTEGER;
$j = 0$;
FOR $k = 1, \ldots, N$ DO
    with probability $\dfrac{n-j}{N-(k-1)}$ THEN $\left| \begin{array}{l} \text{select unit } k; \\ j = j+1; \end{array} \right.$
ENDFOR.

This method was improved by Fan et al. (1962), Knuth (1981), Vitter (1987), Ahrens and Dieter (1985), Bissell (1986), and Pinkham (1987) in order to skip the nonselected units directly. These methods are discussed in Deville et al. (1988).

4.4.5 Sample Reservoir Method

The reservoir method has been proposed by McLeod and Bellhouse (1983) and Vitter (1985) but is also a particular case of the Chao procedure (see Chao, 1982, and Section 6.3.6, page 119). As presented in Algorithm 4.4, at the first step, the first n units are selected in the sample. Next, the sample is updated by examining the last $N-n$ units of the file.

The main interest of Algorithm 4.4 is that the population size must not be known before applying the algorithm. At the end of the file, the population size is known. The reservoir method can be viewed as an eliminatory algorithm. Indeed, at each one of the $N-n$ steps of the algorithm, a unit is definitively eliminated.

Algorithm 4.4 Reservoir method for SRSWOR

DEFINITION k : INTEGER;

The first n units are selected into the sample;

FOR $k = n + 1, \ldots, N$ DO

$\qquad \left|\begin{array}{l} \text{select unit } k; \end{array}\right.$

with probability $\frac{n}{k}$ $\left|\begin{array}{l} \text{a unit is removed from the sample with equal probabilities;} \end{array}\right.$

$\qquad \left|\begin{array}{l} \text{unit } k \text{ takes the place of the removed unit;} \end{array}\right.$

ENDFOR.

Result 12. *(McLeod and Bellhouse, 1983) Algorithm 4.4 gives a SRSWOR.*

Proof. (by induction) Let U_j denote the population of the first j units of U with $j = n, n + 1, \ldots, N$, and

$$\mathcal{S}_n(U_j) = \{ \mathbf{s} \in \mathcal{S}_n \, | s_k = 0, \text{ for } k = j + 1, \ldots, N \} .$$

Let $\mathbf{S}(j)$ denote a random sample of size n selected in support $\mathcal{S}_n(U_j)$. All the vectors

$$\mathbf{s}(j) = (s_1(j) \ \cdots \ s_k(j) \ \cdots \ s_N(j))'$$

of $\mathcal{S}_n(U_j)$ are possible values for the random sample $\mathbf{S}(j)$. We prove that

$$\Pr[\mathbf{S}(j) = \mathbf{s}(j)] = \binom{j}{n}^{-1}, \text{ for } \mathbf{s}(j) \subset \mathcal{S}_n(U_j). \tag{4.3}$$

Expression (4.3) is true when $j = n$. Indeed, in this case, $\mathcal{S}_n(U_j)$ only contains a sample $\mathbf{s}(n) = U_n$ and $\Pr(\mathbf{S}_n = \mathbf{s}_n) = 1$. When $j > n$, Expression (4.3) is proved by induction, supposing that (4.3) is true for $j - 1$. Considering a sample $\mathbf{s}(j)$ of $\mathcal{S}_n(U_j)$, two cases must be distinguished.

- Case 1. $s_j(j) = 0$
  We obtain

$$\Pr[\mathbf{S}(j) = \mathbf{s}(j)]$$
$$= \Pr[\mathbf{S}(j - 1) = \mathbf{s}(j)] \times \Pr(\text{unit } j \text{ is not selected at step } j)$$
$$= \binom{j - 1}{n}^{-1} \left(1 - \frac{n}{j}\right) = \binom{j}{n}^{-1} . \tag{4.4}$$

- Case 2. $s_j(j) = 1$
  In this case, the probability of selecting sample $\mathbf{s}(j)$ at step j is the probability that:
  1. At step $j - 1$, a sample $\mathbf{s}(j) - \mathbf{a}_j + \mathbf{a}_i$ is selected, where i is any unit of U_{j-1} that is not selected in $\mathbf{s}_j$ and $\mathbf{a}_j = (0 \ \cdots \ 0 \ \underbrace{1}_{j\text{th}} \ 0 \ \cdots \ 0)$;
  2. Unit j is selected;
  3. Unit i is replaced by j.

This probability can thus be written

$$\Pr[\mathbf{S}_j = \mathbf{s}(j)] = \sum_{i \in U_{j-1}|s_i(j)=0} \Pr\left[\mathbf{S}(j-1) = \mathbf{s}(j) - \mathbf{a}_j + \mathbf{a}_i\right] \times \frac{n}{j} \times \frac{1}{n}$$

$$= (j-n)\binom{j-1}{n}^{-1}\frac{1}{j} = \binom{j}{n}^{-1}. \tag{4.5}$$

Results (4.4) and (4.5) prove Expression (4.3) by induction. Finally, Result 12 is obtained by taking $j = N$ in Expression (4.3). $\qquad\square$

4.4.6 Random Sorting Method for SRSWOR

A very simple method presented in Algorithm 4.5 consists of randomly sorting the population file.

Algorithm 4.5 Random sort procedure for SRSWOR

---

1. A value of an independent uniform variable in [0,1] is allocated to each unit of the population.
2. The population is sorted in ascending (or descending) order.
3. The first (or last) n units of the sorted population are selected in the sample.

Sunter (1977, p. 273) has proved that the random sorting gives a SRSWOR.

Result 13. *The random sorting method gives a SRSWOR.*

Proof. Consider the distribution function of the largest uniform random number generated for $N - n$ units. If u_i denotes the uniform random variable of the ith unit of the file, this distribution function is given by:

$$F(x) = \Pr\left[\bigcap_{i=1}^{N-n} (u_i \leq x)\right] = \prod_{i=1}^{N-n} \Pr(u_i \leq x) = x^{N-n}, \quad 0 < x < 1.$$

The probability that the n remaining units are all greater than x is equal to $(1-x)^n$. The probability of selecting a particular sample $\mathbf{s}$ is thus

$$p(\mathbf{s}) = \int_0^1 (1-x)^n dF(x) = (N-n)\int_0^1 (1-x)^n x^{N-n-1} dx, \tag{4.6}$$

which is an Euler integral

$$B(x,y) = \int_0^1 t^x (1-t)^y dt = \frac{x!y!}{(x+y+1)!}, \quad x, y \text{ integer.} \tag{4.7}$$

From (4.6) and (4.7), we get

$$p(\mathbf{s}) = \binom{N}{n}^{-1}. \qquad\square$$

4.5 Bernoulli Sampling With Replacement (BERNWR)

4.5.1 Sampling Design

Bernoulli sampling with replacement is not useful in practice. However, we introduce it for the completeness of the definitions.

Definition 43. *A simple design defined on support $\mathcal{R}$ is called Bernoulli sampling design with replacement.*

The Bernoulli sampling design with replacement can be deduced from Definition 40, page 41:

$$p_{\text{BERNWR}}(\mathbf{s}, \mu) = p_{\text{SIMPLE}}(\mathbf{s}, \theta = \mu, \mathcal{R}) = \frac{\mu^{n(\mathbf{s})} \prod_{k \in U} \frac{1}{s_k!}}{\sum_{\mathbf{s} \in \mathcal{R}} \mu^{n(\mathbf{s})} \prod_{k \in U} \frac{1}{s_k!}}$$

$$= \frac{\mu^{n(\mathbf{s})} \prod_{k \in U} \frac{1}{s_k!}}{\prod_{k \in U} \sum_{i=0}^{\infty} \frac{\mu^i}{i!}} = e^{-N\mu} \mu^{n(\mathbf{s})} \prod_{k \in U} \frac{1}{s_k!},$$

for all $\mathbf{s} \in \mathcal{R}$ and with $\mu \in \mathbb{R}_+^*$.

Because we can also write the sampling design as

$$p_{\text{BERNWR}}(\mathbf{s}, \mu) = \prod_{k \in U} \frac{e^{-\mu} \mu^{s_k}}{s_k!},$$

where the $S_1, \ldots, S_N$ are independent and identically distributed Poisson variables with parameter μ, the following results can be derived directly from the Poisson distribution.

The expectation is $\mu_k = \mathrm{E}(S_k) = \mu$, for all $k \in U$. The joint expectation is

$$\mu_{k\ell} = \mathrm{E}(S_k S_\ell) = \mathrm{E}(S_k)\mathrm{E}(S_\ell) = \mu^2, \text{ for all } k \neq \ell \in U.$$

In a Poisson distribution, the variance is equal to the expectation

$$\Sigma_{kk} = \mathrm{var}(S_k) = \mu, \text{ for all } k \in U.$$

Due to the independence of the S_k,

$$\Sigma_{k\ell} = \mathrm{cov}(S_k, S_\ell) = 0, \text{ for all } k \neq \ell \in U.$$

The variance-covariance operator is $\boldsymbol{\Sigma} = \mathbf{I}\mu$. Because the sum of independent Poisson variables is a Poisson variable, the probability distribution of the sample size comes directly:

$$\Pr\left[n(\mathbf{S}) = z\right] = \frac{e^{-(N\mu)}(N\mu)^z}{z!}, \quad z \in \mathbb{N}.$$

The characteristic function is

$$\phi_{\text{BERNWR}}(\mathbf{t}) = \sum_{s \in \mathcal{R}} e^{it's} p_{\text{BERNWR}}(\mathbf{s}, \mu) = \sum_{s \in \mathcal{R}} e^{it's} \prod_{k \in U} \frac{e^{-\mu} \mu^{s_k}}{s_k!}$$

$$= e^{-N\mu} \sum_{s \in \mathcal{R}} \prod_{k \in U} \frac{\left(e^{it_k} \mu\right)^{s_k}}{s_k!} = e^{-N\mu} \prod_{k \in U} \sum_{s_k=0}^{\infty} \frac{\left(e^{it_k} \mu\right)^{s_k}}{s_k!}$$

$$= e^{-N\mu} \prod_{k \in U} \exp\left(e^{it_k} \mu\right) = \prod_{k \in U} \exp\left[\mu\left(e^{it_k} - 1\right)\right] = \exp \mu \sum_{k \in U}(\exp it_k - 1).$$

The inclusion probability is

$$\pi_k = \Pr(S_k > 0) = 1 - \Pr(S_k = 0) = 1 - e^{-\mu}.$$

The joint inclusion probability is

$$\pi_{k\ell} = \Pr(S_k > 0 \text{ and } S_\ell > 0) = [1 - \Pr(S_k = 0)][1 - \Pr(S_\ell = 0)]$$
$$= (1 - e^{-\mu})^2, \quad k \neq \ell.$$

Finally, we have $\boldsymbol{\Delta} = \mathbf{I}(1 - e^{-\mu})e^{-\mu}$, where $\mathbf{I}$ is an $N \times N$ identity matrix.

4.5.2 Estimation

Hansen-Hurwitz estimator

The Hansen-Hurwitz estimator is

$$\widehat{Y}_{HH} = \sum_{k \in U} \frac{y_k S_k}{\mu}.$$

The variance of the Hansen-Hurwitz estimator is

$$\text{var}(\widehat{Y}_{HH}) = \sum_{k \in U} \frac{y_k^2}{\mu}.$$

The estimator of the variance is

$$\widehat{\text{var}}(\widehat{Y}_{HH}) = \sum_{k \in U} \frac{S_k y_k^2}{\mu^2}.$$

Improvement of the Hansen-Hurwitz estimator

Because

$$E(S_k | r(\mathbf{S})) = r(S_k) \frac{\mu}{1 - e^{-\mu}},$$

the improved Hansen-Hurwitz estimator is

$$\widehat{Y}_{IHH} = \sum_{k \in U} \frac{y_k \mathrm{E}(S_k | r(\mathbf{S}))}{\mu} = \sum_{k \in U} \frac{y_k r(S_k)}{\mu} \frac{\mu}{1 - e^{-\mu}} = \sum_{k \in U} \frac{y_k r(S_k)}{1 - e^{-\mu}}.$$

As $\mathrm{var}[r(S_k)] = (1 - e^{-\mu})e^{-\mu}$, the variance of the improved Hansen-Hurwitz estimator is given by

$$\mathrm{var}\left(\widehat{Y}_{IHH}\right) = \sum_{k \in U} \frac{y_k^2 \mathrm{var}[r(S_k)]}{(1 - e^{-\mu})^2} = \sum_{k \in U} y_k^2 \frac{(1 - e^{-\mu})e^{-\mu}}{(1 - e^{-\mu})^2} = \sum_{k \in U} \frac{y_k^2}{e^{\mu} - 1}.$$

The improvement brings an important decrease of the variance with respect to the Hansen-Hurwitz estimator. This variance can be estimated by

$$\widehat{\mathrm{var}}\left(\widehat{Y}_{IHH}\right) = \sum_{k \in U} \frac{r(S_k)y_k^2}{(e^{\mu} - 1)^2}.$$

The Horvitz-Thompson estimator

Because $\pi_k = 1 - e^{-\mu}$, the Horvitz-Thompson estimator is

$$\widehat{Y}_{HT} = \sum_{k \in U} \frac{y_k r(S_k)}{\pi_k} = \sum_{k \in U} \frac{y_k r(S_k)}{1 - e^{-\mu}},$$

which is, in this case, the same estimator as the improved Hansen-Hurwitz estimator.

4.5.3 Sequential Procedure for BERNWR

Due to the independence of the S_k, the standard sequential Algorithm 3.2, page 34, is strictly sequential and simple to implement for a BERNWR, as presented in Algorithm 4.6.

Algorithm 4.6 Bernoulli sampling with replacement

DEFINITION k : INTEGER;
FOR $k = 1, \ldots, N$ DO
    select randomly s_k times unit k according to the Poisson distribution $\mathcal{P}(\mu)$;
ENDFOR.

4.6 Simple Random Sampling With Replacement (SRSWR)

4.6.1 Sampling Design

Definition 44. *A simple design defined on support $\mathcal{R}_n$ is called a simple random sampling with replacement.*

The simple random sampling with replacement can be deduced from Definition 40, page 41:

$$p_{\text{SRSWR}}(\mathbf{s}, n) = p_{\text{SIMPLE}}(\mathbf{s}, \theta, \mathcal{R}_n) = \frac{\theta^{n(\mathbf{s})} \prod_{k \in U} \frac{1}{s_k!}}{\sum_{\mathbf{s} \in \mathcal{R}_n} \theta^{n(\mathbf{s})} \prod_{k \in U} \frac{1}{s_k!}}$$

$$= \frac{n!}{N^n} \prod_{k \in U} \frac{1}{s_k!} = \frac{n!}{s_1! \dots s_N!} \prod_{k \in U} \left(\frac{1}{N}\right)^{s_k},$$

for all $\mathbf{s} \in \mathcal{R}_n$. Note that $p_{\text{SRSWR}}(\mathbf{s}, n)$ does not depend on θ anymore. The sampling design is multinomial.

The characteristic function is

$$\phi_{\text{SRSWR}}(\mathbf{t}) = \sum_{\mathbf{s} \in \mathcal{R}_n} e^{i\mathbf{t}'\mathbf{s}} p_{\text{SRSWR}}(\mathbf{s}, n) = \sum_{\mathbf{s} \in \mathcal{R}_n} e^{i\mathbf{t}'\mathbf{s}} \frac{n!}{N^n} \prod_{k \in U} \frac{1}{s_k!}$$

$$= \sum_{\mathbf{s} \in \mathcal{R}_n} n! \prod_{k \in U} \left(\frac{e^{it_k}}{N}\right)^{s_k} \frac{1}{s_k!} = \left(\frac{1}{N} \sum_{k \in U} \exp it_k\right)^n.$$

The expectation of $\mathbf{S}$ is $\boldsymbol{\mu} = \left(\frac{n}{N} \cdots \frac{n}{N}\right)'$. The joint expectation is

$$\mu_{k\ell} = \mathrm{E}(S_k S_\ell) = \begin{cases} \dfrac{n(N-1+n)}{N^2}, & k = \ell \\ \dfrac{n(n-1)}{N^2}, & k \neq \ell. \end{cases}$$

The variance-covariance operator is

$$\Sigma_{k\ell} = \mathrm{E}(S_k S_\ell) = \begin{cases} \dfrac{n(N-1+n)}{N} - \dfrac{n^2}{N^2} = \dfrac{n(N-1)}{N^2}, & k = \ell \\ \dfrac{n(n-1)}{N^2} - \dfrac{n^2}{N^2} = -\dfrac{n}{N^2}, & k \neq \ell. \end{cases}$$

Thus $\boldsymbol{\Sigma} = \mathrm{var}(\mathbf{S}) = \mathbf{H}n/N$, where $\mathbf{H}$ is defined in (4.1). Moreover,

$$\frac{\Sigma_{k\ell}}{\mu_{k\ell}} = \begin{cases} \dfrac{(N-1)}{(N-1+n)}, & k = \ell \\ -\dfrac{1}{(n-1)}, & k \neq \ell. \end{cases} \tag{4.8}$$

The inclusion probability is

$$\pi_k = \Pr(S_k > 0) = 1 - \Pr(S_k = 0) = 1 - \left(\frac{N-1}{N}\right)^n.$$

The joint inclusion probability is

$$\pi_{k\ell} = \Pr(S_k > 0 \text{ and } S_\ell > 0) = 1 - \Pr(S_k = 0 \text{ or } S_\ell = 0)$$

$$= 1 - \Pr(S_k = 0) - \Pr(S_\ell = 0) + \Pr(S_k = 0 \text{ and } S_\ell = 0)$$

$$= 1 - 2\left(\frac{N-1}{N}\right)^n + \left(\frac{N-2}{N}\right)^n.$$

4.6.2 Distribution of $n[r(\mathbf{S})]$

The question of the distribution of $n[r(\mathbf{S})]$ in simple random sampling with replacement has been studied by Basu (1958), Raj and Khamis (1958), Pathak (1962), Pathak (1988), Chikkagoudar (1966), Konijn (1973, chapter IV), and Cassel et al. (1993, p. 41). In sampling with replacement, it is always preferable to suppress the information about the multiplicity of the units (see Theorem 3, page 24). If we conserve only the distinct units, the sample size becomes random. The first objective is thus to compute the sample size of the distinct units.

Result 14. *If $n[r(\mathbf{S})]$ is the number of distinct units obtained by simple random sampling of m units with replacement in a population of size N, then*

$$\Pr\left\{n[r(\mathbf{S})] = z\right\} = \frac{N!}{(N-z)!N^m}\mathfrak{s}_m^{(z)}, \quad z = 1, \ldots, \min(m, N), \qquad (4.9)$$

where $\mathfrak{s}_m^{(z)}$ is a Stirling number of the second kind

$$\mathfrak{s}_m^{(z)} = \frac{1}{z!}\sum_{i=1}^{z}\binom{z}{i}i^m(-1)^{z-i}.$$

Proof. (by induction) The Stirling numbers of the second kind satisfy the recursive relation:

$$\mathfrak{s}_m^{(z)} = \mathfrak{s}_{m-1}^{(z-1)} + z\mathfrak{s}_{m-1}^{(z)}, \qquad (4.10)$$

with the initial values $\mathfrak{s}_1^{(1)} = 1$ and $\mathfrak{s}_1^{(z)} = 0, z \neq 1$ (see Abramowitz and Stegun, 1964, pp. 824-825). If $n[r(\mathbf{S}_m)]$ is the number of distinct units obtained by simple random sampling with replacement of $m-1$ units in a population of size N, then we have:

$$\begin{aligned}
&\Pr\left\{n[r(\mathbf{S}_m)] = z\right\} \\
&= \Pr\left\{n[r(\mathbf{S}_{m-1})] = z-1\right\} \\
&\quad \times \Pr\left\{\text{select at the } m\text{th draw a unit not yet selected}\right. \\
&\qquad \left. \text{given that } n[r(\mathbf{S}_{m-1})] = z-1\right\} \\
&\quad + \Pr\left\{n[r(\mathbf{S}_{m-1})] = z\right\} \\
&\quad \times \Pr\left\{\text{select at the } m\text{th draw an unit already selected}\right. \\
&\qquad \left. \text{given that } n[r(\mathbf{S}_{m-1})] = z\right\} \\
&= \Pr\left\{n[r(\mathbf{S}_{m-1})] = z-1\right\}\frac{N-z+1}{N} + \Pr\left\{n[r(\mathbf{S}_{m-1})] = z\right\}\frac{z}{N}.
\end{aligned}$$

$$(4.11)$$

Moreover, we have the initial conditions,

$$\Pr\left\{n[r(\mathbf{S}_1)] = z\right\} = \begin{cases} 1 & \text{if } z = 1 \\ 0 & \text{if } z \neq 1. \end{cases}$$

If we suppose that Result 14 is true at step $m - 1$, and by property (4.10), the recurrence equation is satisfied by Expression (4.9). Indeed

$$
\Pr\left\{n[r(\mathbf{S}_{m-1})] = z - 1]\right\} \frac{N - z + 1}{N} + \Pr\left\{n[r(\mathbf{S}_{m-1})] = z\right\} \frac{z}{N}
$$
$$
= \frac{N!}{(N - z + 1)! N^{m-1}} \mathbf{S}_{m-1}^{(z-1)} \frac{N - z + 1}{N} + \frac{N!}{(N - z)! N^{m-1}} \mathbf{S}_{m-1}^{(z)} \frac{z}{N}
$$
$$
= \frac{N!}{(N - z)! N^m} \left[\mathbf{S}_{m-1}^{(z-1)} + z \mathbf{S}_{m-1}^{(z)} \right] = \frac{N!}{(N - z)! N^m} \mathbf{S}_m^{(z)} = \Pr\left\{n[r(\mathbf{S}_m)] = z\right\}.
$$

As the initial conditions are also satisfied, Result 14 is proved. $\square$

The following result is used to derive the mean and variance of $n[r(\mathbf{S})]$.

Result 15. *If $n[r(\mathbf{S})]$ is the number of distinct units by selecting m units with replacement in a population of size N, then*

$$
\mathrm{E}\left[\prod_{i=0}^{j-1} (N - n[r(\mathbf{S})] - i)\right] = \frac{(N - j)^m N!}{N^m (N - j)!}, j = 1, \ldots, N. \tag{4.12}
$$

Proof.

$$
\mathrm{E}\left(\prod_{i=0}^{j-1} \{N - n[r(\mathbf{S})] - i\}\right)
$$
$$
= \sum_{z=1}^{\min(m,N)} \left[\prod_{i=0}^{j-1} (N - z - i)\right] \frac{N!}{(N - z)! N^m} \mathbf{S}_m^{(z)}
$$
$$
= \sum_{z=1}^{\min(m,N-j)} \frac{(N - z)!}{(N - z - j)!} \frac{N!}{(N - z)! N^m} \mathbf{S}_m^{(z)}
$$
$$
= \frac{(N - j)^m N!}{N^m (N - j)!} \sum_{z=1}^{\min(m,N-j)} \frac{(N - j)!}{(N - z - j)!(N - j)^m} \mathbf{S}_m^{(z)}. \tag{4.13}
$$

Because

$$
\frac{(N - j)!}{(N - z - j)!(N - j)^m} \mathbf{S}_m^{(z)}
$$

is the probability of obtaining exactly z distinct units by selecting m units in a population of size $N - j$, we obtain

$$
\sum_{z=1}^{\min(m,N-j)} \frac{(N - j)!}{(N - z - j)!(N - j)^m} \mathbf{S}_m^{(z)} = 1,
$$

and by (4.13), we get directly (4.12). $\square$

By (4.12) when $j = 1$, the expectation can be derived

$$E\{n[r(\mathbf{S})]\} = N\left[1 - \left(\frac{N-1}{N}\right)^m\right]. \tag{4.14}$$

By (4.12) when $j = 2$, after some algebra, the variance can be obtained

$$\text{var}\{n[r(\mathbf{S})]\} = \frac{(N-1)^m}{N^{m-1}} + (N-1)\frac{(N-2)^m}{N^{m-1}} - \frac{(N-1)^{2m}}{N^{2m-2}}. \tag{4.15}$$

4.6.3 Estimation

Hansen-Hurwitz estimator

The Hansen-Hurwitz estimator is

$$\widehat{Y}_{HH} = \frac{N}{n}\sum_{k\in U} y_k S_k,$$

and its variance given in (2.12) and (2.13) is

$$\text{var}(\widehat{Y}_{HH}) = \frac{1}{n}\sum_{k\in U} y_k^2 - \frac{1}{n}\left(\sum_{k\in U} y_k\right)^2 = \frac{N^2}{n}\frac{1}{N}\sum_{k\in U}\left(y_k - \frac{1}{N}\sum_{k\in U} y_k\right)^2$$

$$= \frac{N^2\sigma_y^2}{n}, \tag{4.16}$$

where

$$\sigma_y^2 = \frac{1}{N}\sum_{k\in U}\left(y_k - \bar{Y}\right)^2 = \frac{1}{2N^2}\sum_{k\in U}\sum_{\ell\in U}(y_k - y_\ell)^2.$$

For estimating the variance, we have the following result.

Result 16. *In a SRSWR, v_y^2 is an unbiased estimator of σ_y^2, where*

$$v_y^2 = \frac{1}{n-1}\sum_{k\in U} S_k\left(y_k - \frac{\widehat{Y}_{HH}}{N}\right)^2.$$

Proof. Because

$$v_y^2 = \frac{1}{n-1}\sum_{k\in U} S_k\left(y_k - \frac{\widehat{Y}_{HH}}{N}\right)^2 = \frac{1}{2n(n-1)}\sum_{k\in U}\sum_{\ell\in U}(y_k - y_\ell)^2 S_k S_\ell,$$

we obtain

$$E(v_y^2) = \frac{1}{2n(n-1)}\sum_{k\in U}\sum_{\ell\in U}(y_k - y_\ell)^2 \, E\left(S_k S_\ell\right)$$

$$= \frac{1}{2n(n-1)}\sum_{k\in U}\sum_{\ell\in U}(y_k - y_\ell)^2 \frac{n(n-1)}{N^2}$$

$$= \frac{1}{2N^2}\sum_{k\in U}\sum_{\ell\in U}(y_k - y_\ell)^2 = \sigma_y^2. \qquad \square$$

Note that depending on the sampling design being a SRSWR or a SR-SWOR, the sample variance v_y^2 estimates distinct functions of interest (see Result 16, page 57 and Result 11, page 47).

By Result 16 and Expression (4.16), we have an unbiased estimator of the variance:

$$\widehat{\mathrm{var}}_2(\widehat{Y}_{HH}) = \frac{N^2 v_y^2}{n}. \tag{4.17}$$

Estimator (4.17) is also a particular case of the estimator given in Expression (2.15), page 27.

The other unbiased estimator given in Expression (2.14), page 27, provides a very strange expression. Indeed, by Result 9, page 27 and Expression (4.8), page 54, we have

$$\widehat{\mathrm{var}}_1(\widehat{Y}_{HH}) = \widehat{\mathrm{var}}_2(\widehat{Y}_{HH}) + \sum_{k \in U} \frac{S_k y_k^2}{\mu_k^2} \sum_{\ell \in U} \frac{S_\ell \Sigma_{k\ell}}{\mu_{k\ell}}$$

$$= \frac{N^2 v_y^2}{n} + \sum_{k \in U} \frac{y_k^2}{\mu_k^2} \left[S_k^2 \frac{Nn}{(n-1)(N-1+n)} - S_k \frac{n}{n-1} \right].$$

Although unbiased, $\widehat{\mathrm{var}}_1(\widehat{Y}_{HH})$ should never be used.

Improved Hansen-Hurwitz estimator

Because

$$\mathrm{E}\left[S_k | r(\mathbf{S})\right] = r(S_k) \frac{n}{n[r(\mathbf{S})]},$$

the improved Hansen-Hurwitz estimator is

$$\widehat{Y}_{IHH} = \frac{N}{n} \sum_{k \in U} y_k \mathrm{E}\left[S_k | r(\mathbf{S})\right] = \frac{N}{n[r(\mathbf{S})]} \sum_{k \in U} y_k r(S_k),$$

and its variance is

$$\mathrm{var}(\widehat{Y}_{IHH}) = \mathrm{E}\,\mathrm{var}\left\{ \frac{N}{n[r(\mathbf{S})]} \sum_{k \in U} y_k r(S_k) \middle| n[r(\mathbf{S})] \right\}$$

$$+ \mathrm{var}\,\mathrm{E}\left\{ \frac{N}{n[r(\mathbf{S})]} \sum_{k \in U} y_k r(S_k) \middle| n[r(\mathbf{S})] \right\}.$$

The improved Hansen-Hurwitz estimator is unbiased conditionally on $n[r(\mathbf{S})]$, thus the second term vanishes. Conditionally on $n[r(\mathbf{S})]$, the sample is simple with fixed sample size, thus

$$\mathrm{var}(\widehat{Y}_{IHH}) = \mathrm{E}\,\mathrm{var}\left\{ \frac{N}{n[r(\mathbf{S})]} \sum_{k \in U} y_k r(S_k) \middle| n[r(\mathbf{S})] \right\}$$

$$= \mathrm{E}\left\{ N^2 \frac{N - n[r(\mathbf{S})]}{N} \frac{V_y^2}{n[r(\mathbf{S})]} \right\} = N^2 \left(\mathrm{E}\left\{ \frac{1}{n[r(\mathbf{S})]} \right\} - \frac{1}{N} \right) V_y^2.$$

In order to derive the variance, we have to compute $E\left[1/n(\mathbf{S})\right]$.

Result 17. *If $n[r(\mathbf{S})]$ is the number of distinct units obtained by selecting m units according to a SRSWR, then*

$$E\left\{\frac{1}{n[r(\mathbf{S})]}\right\} = \frac{1}{N^m}\sum_{j=1}^{N} j^{m-1}. \qquad (4.18)$$

Proof. (by induction) Let $n[r(\mathbf{S}), N]$ be the number of distinct units obtained by selecting m units according to a SRSWR from a population of size N. We first show that

$$E\left\{\frac{N^m}{n[r(\mathbf{S}), N]}\right\} - E\left\{\frac{(N-1)^m}{n[r(\mathbf{S}), N-1]}\right\} = N^{m-1}. \qquad (4.19)$$

Indeed,

$$E\left[\frac{N^m}{n[r(\mathbf{S}), N]} - \frac{(N-1)^m}{n[r(\mathbf{S}), N-1]}\right]$$

$$= \sum_{z=1}^{\min(m,N)} \frac{1}{z}\frac{N!}{(N-z)!}\mathfrak{s}_m^{(z)} - \sum_{z=1}^{\min(m,N-1)} \frac{1}{z}\frac{(N-1)!}{(N-1-z)!}\mathfrak{s}_m^{(z)}$$

$$= N^{m-1}\sum_{z=1}^{\min(m,N)} \frac{N!}{(N-z)!N^m}\mathfrak{s}_m^{(z)} = N^{m-1}.$$

By (4.19), we get a recurrence equation

$$E\left\{\frac{1}{n[r(\mathbf{S}), N]}\right\} = \frac{1}{N} + \frac{(N-1)^m}{N^m}E\left\{\frac{1}{n[r(\mathbf{S}), N-1]}\right\}.$$

The initial condition is obvious:

$$E\left\{\frac{1}{n[r(\mathbf{S}), 1]}\right\} = 1.$$

Expression (4.18) satisfies the recurrence equation and the initial condition.
□

The variance is thus:

$$\mathrm{var}(\widehat{Y}_{IHH}) = N^2\left(\frac{1}{N^m}\sum_{j=1}^{N} j^{m-1} - \frac{1}{N}\right)V_y^2 = \frac{V_y^2}{N^{m-2}}\sum_{j=1}^{N-1} j^{m-1}.$$

The Horvitz-Thompson estimator

The Horvitz-Thompson estimator is

$$\widehat{Y}_{HT} = \left[1 - \left(\frac{N}{N-1} \right)^n \right]^{-1} \sum_{k \in U} y_k r(S_k).$$

The variance is thus

$$\mathrm{var}\left(\widehat{Y}_{HT} \right)$$

$$= \mathrm{E}\,\mathrm{var}\left\{ \widehat{Y}_{HT} | n[r(\mathbf{S})] \right\} + \mathrm{var}\,\mathrm{E}\left\{ \widehat{Y}_{HT} | n[r(\mathbf{S})] \right\}$$

$$= \frac{1}{\pi^2}\mathrm{E}\,\mathrm{var}\left\{ n[r(\mathbf{S})]\widehat{Y}_{IHH} | n[r(\mathbf{S})] \right\} + \frac{1}{\pi^2}\mathrm{var}\,\mathrm{E}\left\{ n[r(\mathbf{S})]\widehat{Y}_{IHH} | n[r(\mathbf{S})] \right\}$$

$$= \frac{1}{\pi^2}\mathrm{E}\left(\{n[r(\mathbf{S})]\}^2 \,\mathrm{var}\left\{ \widehat{Y}_{IHH} | n[r(\mathbf{S})] \right\} \right) + \frac{Y^2}{\pi^2}\mathrm{var}\left\{ n[r(\mathbf{S})] \right\}$$

$$= \frac{1}{\pi^2}\mathrm{E}\left(\{n[r(\mathbf{S})]\}^2 \, N^2 \frac{N - n[r(\mathbf{S})]}{N} \frac{V_y^2}{n[r(\mathbf{S})]} \right) + \frac{Y^2}{\pi^2}\mathrm{var}\left\{ n[r(\mathbf{S})] \right\}$$

$$= \frac{NV_y^2}{\pi^2}\mathrm{E}\left\{ Nn[r(\mathbf{S})] - n[r(\mathbf{S})]^2 \right\} + \frac{Y^2}{\pi^2}\mathrm{var}\left[n[r(\mathbf{S})] \right]$$

$$= \frac{NV_y^2}{\pi^2}\left[N\mathrm{E}\left\{ n[r(\mathbf{S})] \right\} - \mathrm{var}\left\{ n[r(\mathbf{S})]^2 \right\} + \left(\mathrm{E}\left\{ n[r(\mathbf{S})] \right\} \right)^2 \right] + \frac{Y^2}{\pi^2}\mathrm{var}\left\{ n[r(\mathbf{S})] \right\}.$$

This variance is quite complex. Estimator $\widehat{Y}_{IHH}$ does not necessarily have a smaller variance than $\widehat{Y}_{HT}$. However, $\widehat{Y}_{HT}$ should not be used because it depends directly on Y^2; that means that even if the y_k's are constant, $\mathrm{var}(\widehat{Y}_{HT})$ is not equal to zero. We thus advocate for the use of $\widehat{Y}_{IHH}$.

4.6.4 Draw by Draw Procedure for SRSWR

The standard draw by draw Algorithm 3.3, page 35, gives a well-known procedure for simple random sampling: at each of the N steps of the algorithm, a unit is selected from the population with equal probability (see Example 5, page 36). This method is summarized in Algorithm 4.7.

Algorithm 4.7 Draw by draw procedure for SRSWR

DEFINITION j : INTEGER;
FOR $j = 1, \ldots, n$ DO
    a unit is selected with equal probability $1/N$ from the population U;
ENDFOR.

4.6.5 Sequential Procedure for SRSWR

As SRSWOR has a fixed sample size, a standard sequential Algorithm 3.2, page 34, can be easily implemented. This neglected method, presented in Algorithm 4.8, is better than the previous one because it allows selecting the sample in one pass.

Algorithm 4.8 Sequential procedure for SRSWR

DEFINITION k, j : INTEGER;

$j = 0$;

FOR $k = 1, \ldots, N$ DO

    select the kth unit s_k times according to the binomial distribution

$$\mathcal{B}\left(n - \sum_{i=1}^{k-1} s_i, \frac{1}{N-k+1} \right);$$

ENDFOR.

4.7 Links Between the Simple Sampling Designs

The following relations of conditioning can be proved using conditioning with respect to a sampling design:

- $p_{\text{BERNWR}}(\mathbf{s}, \mu | \mathcal{R}_n) = p_{\text{SRSWR}}(\mathbf{s}, n)$, for all $\mu \in \mathbb{R}_+^*$,

- $p_{\text{BERNWR}}(\mathbf{s}, \theta | \mathcal{S}) = p_{\text{BERN}}\left(\mathbf{s}, \pi = \dfrac{\theta}{1+\theta} \right)$, for all $\theta \in \mathbb{R}_+^*$,

- $p_{\text{BERNWR}}(\mathbf{s}, \theta | \mathcal{S}_n) = p_{\text{SRSWOR}}(\mathbf{s}, n)$, for all $\theta \in \mathbb{R}_+^*$,

- $p_{\text{SRSWR}}(\mathbf{s}, n | \mathcal{S}_n) = p_{\text{SRSWOR}}(\mathbf{s}, n)$,

- $p_{\text{BERN}}(\mathbf{s}, \pi | \mathcal{S}_n) = p_{\text{SRSWOR}}(\mathbf{s}, n)$, for all $\pi \in]0, 1[$.

These relations are summarized in Figure 4.1 and Table 4.1, page 62.

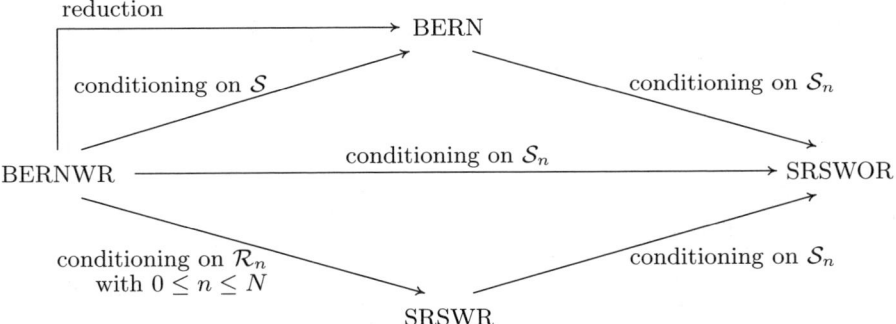

Fig. 4.1. Links between the main simple sampling designs

Table 4.1. Main simple random sampling designs

	Notation	BERNWR	SRSWR	BERN	SRSWOR
Design	$p(\mathbf{s})$	$\dfrac{\mu^{n(s)}}{e^{N\mu}}\prod\limits_{k\in U}\dfrac{1}{s_k!}$	$\dfrac{n!}{N^n}\prod\limits_{k\in U}\dfrac{1}{s_k!}$	$\pi^{n(\mathbf{s})}(1-\pi)^{N-n(\mathbf{s})}$	$\dbinom{N}{n}^{-1}$
Support	Q	$\mathcal{R}$		$\mathcal{S}$	$\mathcal{S}_n$
Char. function	$\phi(\mathbf{t})$	$\exp\left\{\mu\sum\limits_{k\in U}(e^{it_k}-1)\right\}$	$\left(\dfrac{1}{N}\sum\limits_{k\in U}e^{it_k}\right)^n$	$\prod\limits_{k\in U}\{1+\pi(e^{it_k}-1)\}$	$\dbinom{N}{n}^{-1}\sum\limits_{s\in S_n}e^{it's}$
Replacement	WOR/WR	with repl.	with repl.	without repl.	without repl.
Sample size	$n(\mathbf{S})$	random	fixed	random	fixed
Expectation	μ_k	μ	$\dfrac{n}{N}$	π	$\dfrac{n}{N}$
Inclusion probability	π_k	$1-e^{-\mu}$	$1-\left(\dfrac{N-1}{N}\right)^n$	π	$\dfrac{n}{N}$
Variance	Σ_{kk}	μ	$\dfrac{n(N-1)}{N^2}$	$\pi(1-\pi)$	$\dfrac{n(N-n)}{N^2}$
Joint expectation	$\mu_{k\ell}, k\neq\ell$	μ^2	$\dfrac{n(n-1)}{N^2}$	π^2	$\dfrac{n(n-1)}{N(N-1)}$
Covariance	$\Sigma_{k\ell}, k\neq\ell$	0	$-\dfrac{n}{N^2}$	0	$-\dfrac{n(N-n)}{N^2(N-1)}$
Basic est.	$\hat{Y}$	$\sum\limits_{k\in U}\dfrac{y_k S_k}{\mu}$	$\dfrac{N}{n}\sum\limits_{k\in U}y_k S_k$	$\sum\limits_{k\in U}\dfrac{y_k S_k}{\pi}$	$\dfrac{N}{n}\sum\limits_{k\in U}y_k S_k$
Variance	$\mathrm{var}(\hat{Y})$	$\sum\limits_{k\in U}\dfrac{y_k^2}{\mu}$	$\dfrac{N^2\sigma_y^2}{n}$	$\dfrac{1-\pi}{\pi}\sum\limits_{k\in U}y_k^2$	$N^2\dfrac{N-n}{N}\dfrac{V_y^2}{n}$
Variance est.	$\widehat{\mathrm{var}}(\hat{Y})$	$\sum\limits_{k\in U}\dfrac{y_k^2}{\mu^2}S_k$	$N^2\dfrac{v_y^2}{n}$	$(1-\pi)\sum\limits_{k\in U}y_k^2 S_k$	$N^2\dfrac{N-n}{N}\dfrac{v_y^2}{n}$

5

Unequal Probability Exponential Designs

5.1 Introduction

A sampling design is a multivariate discrete distribution and an exponential design is thus an exponential multivariate discrete distribution. Exponential designs are a large family that includes simple random sampling, multinomial sampling, Poisson sampling, unequal probability sampling with replacement, and conditional or rejective Poisson sampling. A large part of the posthumous book of Hájek (1981) is dedicated to the exponential family. Hájek advocated for the use of Poisson rejective sampling but the link between the inclusion probabilities of Poisson sampling and conditional Poisson sampling was not yet clearly elucidated. Chen et al. (1994) have taken a big step forward by linking the question of conditional Poisson sampling to the general theory of the exponential family.

The fundamental point developed by Chen et al. (1994) is the link between the parameter of the exponential family and its mean; that is, its vector of inclusion probabilities. Once this link is clarified, the implementation of the classical algorithms follows quite easily. Independently, Jonasson and Nerman (1996), Aires (1999, 2000a), Bondesson et al. (2004), and Traat et al. (2004) have investigated the question of conditional Poisson sampling. Chen and Liu (1997), Chen (1998), and Deville (2000) have improved the algorithms and the technique of computation of inclusion probabilities. A large part of the material of this chapter has been developed during informal conversations with Jean-Claude Deville. In this chapter, we attempt to present a coherent theory of exponential designs. A unique definition is given, and all exponential designs can be derived by changes of support. The application of the classical algorithms presented in Chapter 3 allows for the deduction of nine sampling algorithms.

5.2 General Exponential Designs

5.2.1 Minimum Kullback-Leibler Divergence

In order to identify a unique sampling design that satisfies a predetermined vector of means $\boldsymbol{\mu}$, a general idea is to minimize the Kullback-Leibler divergence (see Kullback, 1959):

$$H(p, p_r) = \sum_{\mathbf{s} \in \mathcal{Q}} p(\mathbf{s}) \log \frac{p(\mathbf{s})}{p_r(\mathbf{s})},$$

where $p_r(\mathbf{s})$ is a design of reference on $\mathcal{Q}$. The divergence $H(p, p_r)$ (also called relative entropy) is always positive and $H(p, p_r) = 0$ when $p(\mathbf{s}) = p_r(\mathbf{s})$, for all $\mathbf{s} \in \mathcal{Q}$. The objective is thus to identify the closest sampling design (in the sense of the Kullback-Leibler divergence) to $p_r(\mathbf{s})$ that satisfies fixed inclusion probabilities.

In this chapter, the study of minimum Kullback-Leibler divergence designs is restricted to the case where the reference sampling design is simple; that is,

$$p_r(\mathbf{s}) = p_{\text{SIMPLE}}(\mathbf{s}, \theta, \mathcal{Q}) = \frac{\theta^{n(\mathbf{s})} \prod_{k \in U} \frac{1}{s_k!}}{\sum_{\mathbf{s} \in \mathcal{Q}} \theta^{n(\mathbf{s})} \prod_{k \in U} \frac{1}{s_k!}}. \tag{5.1}$$

Because $H(p, p_r)$ is strictly convex with respect to p, the minimization of $H(p, p_r)$ under linear constraints provides a unique solution. The Kullback-Leibler divergence $H(p, p_r)$ is minimized under the constraints

$$\sum_{\mathbf{s} \in \mathcal{Q}} p(\mathbf{s}) = 1, \tag{5.2}$$

and

$$\sum_{\mathbf{s} \in \mathcal{Q}} \mathbf{s} p(\mathbf{s}) = \boldsymbol{\mu}. \tag{5.3}$$

The vector $\boldsymbol{\mu}$ belongs to $\overset{\circ}{\mathcal{Q}}$ (the interior of $\mathcal{Q}$) (see Remark 4, page 15), which ensures that there exists at least one sampling design on $\mathcal{Q}$ with mean $\boldsymbol{\mu}$.

The Lagrangian function is

$$\mathcal{L}(p(\mathbf{s}), \beta, \boldsymbol{\gamma}) = \sum_{\mathbf{s} \in \mathcal{Q}} p(\mathbf{s}) \log \frac{p(\mathbf{s})}{p_r(\mathbf{s})} - \beta \left(\sum_{\mathbf{s} \in \mathcal{Q}} p(\mathbf{s}) - 1 \right) - \boldsymbol{\gamma}' \left(\sum_{\mathbf{s} \in \mathcal{Q}} \mathbf{s} p(\mathbf{s}) - \boldsymbol{\mu} \right),$$

where $\beta \in \mathbb{R}$ and $\boldsymbol{\gamma} \in \mathbb{R}^N$ are the Lagrange multipliers. By setting the derivatives of $\mathcal{L}$ with respect to $p(\mathbf{s})$ equal to zero, we obtain

$$\frac{\partial \mathcal{L}(p(\mathbf{s}), \beta, \boldsymbol{\gamma})}{\partial p(\mathbf{s})} = \log \frac{p(\mathbf{s})}{p_r(\mathbf{s})} + 1 - \beta - \boldsymbol{\gamma}' \mathbf{s} = 0,$$

which gives
$$p(\mathbf{s}) = p_r(\mathbf{s}) \exp(\boldsymbol{\gamma}'\mathbf{s} + \beta - 1).$$

By using the constraint (5.2), we get

$$p(\mathbf{s}) = \frac{p_r(\mathbf{s}) \exp \boldsymbol{\gamma}'\mathbf{s}}{\sum_{\mathbf{s}\in\mathcal{Q}} p_r(\mathbf{s}) \exp \boldsymbol{\gamma}'\mathbf{s}}.$$

Now, by replacing $p_r(\mathbf{s})$ by (5.1) we get the exponential designs

$$p_{\mathrm{EXP}}(\mathbf{s}, \boldsymbol{\lambda}, \mathcal{Q}) = \frac{\theta^{n(\mathbf{s})} \left(\prod_{k\in U} \frac{1}{s_k!}\right) \exp \boldsymbol{\gamma}'\mathbf{s}}{\sum_{\mathbf{s}\in\mathcal{Q}} \theta^{n(\mathbf{s})} \left(\prod_{k\in U} \frac{1}{s_k!}\right) \exp \boldsymbol{\gamma}'\mathbf{s}} = \frac{\left(\prod_{k\in U} \frac{1}{s_k!}\right) \exp \boldsymbol{\lambda}'\mathbf{s}}{\sum_{\mathbf{s}\in\mathcal{Q}} \left(\prod_{k\in U} \frac{1}{s_k!}\right) \exp \boldsymbol{\lambda}'\mathbf{s}},$$

where $\boldsymbol{\lambda} = (\lambda_1 \cdots \lambda_N)' = (\lambda_1 + \log\theta \cdots \lambda_N + \log\theta)'$. The vector $\boldsymbol{\lambda}$ is identified by means of the constraint (5.3). We show that the problem of deriving $\boldsymbol{\lambda}$ from $\boldsymbol{\mu}$ is one of the most intricate questions of exponential designs.

5.2.2 Exponential Designs (EXP)

We refer to the following definition.

Definition 45. *A sampling design* $p_{\mathrm{EXP}}(.)$ *on a support* $\mathcal{Q}$ *is said to be exponential if it can be written*

$$p_{\mathrm{EXP}}(\mathbf{s}, \boldsymbol{\lambda}, \mathcal{Q}) = g(\mathbf{s}) \exp\left[\boldsymbol{\lambda}'\mathbf{s} - \alpha(\boldsymbol{\lambda}, \mathcal{Q})\right],$$

where $\boldsymbol{\lambda} \in \mathbb{R}^N$ *is the parameter,*

$$g(\mathbf{s}) = \prod_{k\in U} \frac{1}{s_k!},$$

and $\alpha(\boldsymbol{\lambda}, \mathcal{Q})$ *is called the normalizing constant and is given by*

$$\alpha(\boldsymbol{\lambda}, \mathcal{Q}) = \log \sum_{\mathbf{s}\in\mathcal{Q}} g(\mathbf{s}) \exp \boldsymbol{\lambda}'\mathbf{s}.$$

As it is the case for all of the exponential families, the expectation can be obtained by differentiating the normalizing constant

$$\alpha'(\boldsymbol{\lambda}, \mathcal{Q}) = \frac{\partial\alpha(\boldsymbol{\lambda}, \mathcal{Q})}{\partial\boldsymbol{\lambda}} = \frac{\sum_{\mathbf{s}\in\mathcal{Q}} \mathbf{s}g(\mathbf{s}) \exp \boldsymbol{\lambda}'\mathbf{s}}{\sum_{\mathbf{s}\in\mathcal{Q}} g(\mathbf{s}) \exp \boldsymbol{\lambda}'\mathbf{s}} = \boldsymbol{\mu}.$$

By differentiating the normalizing constant twice, we get the variance-covariance operator:

$$\alpha''(\boldsymbol{\lambda}, \mathcal{Q})$$
$$= \frac{\sum_{\mathbf{s}\in\mathcal{Q}} \mathbf{s}\mathbf{s}'g(\mathbf{s}) \exp \boldsymbol{\lambda}'\mathbf{s}}{\sum_{\mathbf{s}\in\mathcal{Q}} g(\mathbf{s}) \exp \boldsymbol{\lambda}'\mathbf{s}} - \left[\frac{\sum_{\mathbf{s}\in\mathcal{Q}} \mathbf{s}g(\mathbf{s}) \exp \boldsymbol{\lambda}'\mathbf{s}}{\sum_{\mathbf{s}\in\mathcal{Q}} g(\mathbf{s}) \exp \boldsymbol{\lambda}'\mathbf{s}}\right]\left[\frac{\sum_{\mathbf{s}\in\mathcal{Q}} \mathbf{s}g(\mathbf{s}) \exp \boldsymbol{\lambda}'\mathbf{s}}{\sum_{\mathbf{s}\in\mathcal{Q}} g(\mathbf{s}) \exp \boldsymbol{\lambda}'\mathbf{s}}\right]'$$
$$= \boldsymbol{\Sigma}.$$

The characteristic function of an exponential design is given by

$$\phi_{\text{EXP}}(\mathbf{t}) = \sum_{\mathbf{s} \in \mathcal{Q}} g(\mathbf{s}) \left[\exp(i\mathbf{t} + \boldsymbol{\lambda}')\mathbf{s} - \alpha(\boldsymbol{\lambda}, \mathcal{Q}) \right] = \exp \left[\alpha(i\mathbf{t} + \boldsymbol{\lambda}, \mathcal{Q}) - \alpha(\boldsymbol{\lambda}, \mathcal{Q}) \right].$$

Simple random sampling designs are a particular case of exponential designs. Indeed, we have

$$p_{\text{EXP}}(\mathbf{s}, \mathbf{1} \log \theta, \mathcal{Q}) = p_{\text{SIMPLE}}(\mathbf{s}, \theta, \mathcal{Q}), \theta \in \mathbb{R}_+,$$

where $\mathbf{1}$ is a vector of N ones.

The parameter $\boldsymbol{\lambda}$ can be split into two parts: $\boldsymbol{\lambda}_a$, that is, the orthogonal projection of $\boldsymbol{\lambda}$ onto $\overrightarrow{\mathcal{Q}}$ (the direction of $\mathcal{Q}$, see Definition 9, page 12) and $\boldsymbol{\lambda}_b$, that is, the orthogonal projection of $\boldsymbol{\lambda}$ onto Invariant $\mathcal{Q}$ (see Definition 11, page 12). We thus have that $\boldsymbol{\lambda} = \boldsymbol{\lambda}_a + \boldsymbol{\lambda}_b$, and $\boldsymbol{\lambda}_a' \boldsymbol{\lambda}_b = 0$. Moreover, we have the following result:

Result 18. *Let $\boldsymbol{\lambda}_a$ and $\boldsymbol{\lambda}_b$ be, respectively, the orthogonal projection on $\overrightarrow{\mathcal{Q}}$ and* Invariant $\mathcal{Q}$. *Then*

$$p_{\text{EXP}}(\mathbf{s}, \boldsymbol{\lambda}, \mathcal{Q}) = p_{\text{EXP}}(\mathbf{s}, \boldsymbol{\lambda}_a + \boldsymbol{\lambda}_b, \mathcal{Q}) = p_{\text{EXP}}(\mathbf{s}, \boldsymbol{\lambda}_a, \mathcal{Q}) = p_{\text{EXP}}(\mathbf{s}, \boldsymbol{\lambda}_a + b\boldsymbol{\lambda}_b, \mathcal{Q}),$$

for any $b \in \mathbb{R}$.

Proof.

$$p_{\text{EXP}}(\mathbf{s}, \boldsymbol{\lambda}_a + b\boldsymbol{\lambda}_b, \mathcal{Q})$$
$$= \frac{g(\mathbf{s}) \exp(\boldsymbol{\lambda}_a + b\boldsymbol{\lambda}_b)'\mathbf{s}}{\sum_{\mathbf{s} \in \mathcal{Q}} g(\mathbf{s}) \exp(\boldsymbol{\lambda}_a + b\boldsymbol{\lambda}_b)'\mathbf{s}} = \frac{g(\mathbf{s})(\exp \boldsymbol{\lambda}_a'\mathbf{s})(\exp b\boldsymbol{\lambda}_b'\mathbf{s})}{\sum_{\mathbf{s} \in \mathcal{Q}} g(\mathbf{s})(\exp \boldsymbol{\lambda}_a'\mathbf{s})(\exp b\boldsymbol{\lambda}_b'\mathbf{s})}.$$

Because $\boldsymbol{\lambda}_b \in$ Invariant $(\mathcal{Q})$, $(\mathbf{s} - \boldsymbol{\mu})'\boldsymbol{\lambda}_b = 0$, and thus $\boldsymbol{\lambda}_b'\mathbf{s} = c$, where $c = \boldsymbol{\mu}'\boldsymbol{\lambda}_b$ for all $\mathbf{s} \in \mathcal{Q}$. Thus,

$$p_{\text{EXP}}(\mathbf{s}, \boldsymbol{\lambda}_a + b\boldsymbol{\lambda}_b, \mathcal{Q})$$
$$= \frac{g(\mathbf{s})(\exp \boldsymbol{\lambda}_a'\mathbf{s})(\exp bc)}{\sum_{\mathbf{s} \in \mathcal{Q}} g(\mathbf{s})(\exp \boldsymbol{\lambda}_a'\mathbf{s})(\exp bc)} = \frac{g(\mathbf{s})(\exp \boldsymbol{\lambda}_a'\mathbf{s})}{\sum_{\mathbf{s} \in \mathcal{Q}} g(\mathbf{s})(\exp \boldsymbol{\lambda}_a'\mathbf{s})} = p_{\text{EXP}}(\mathbf{s}, \boldsymbol{\lambda}_a, \mathcal{Q}). \quad \square$$

Example 8. If the support is $\mathcal{R}_n = \left\{ \mathbf{s} \in \mathbb{N}^N \left| \sum_{k \in U} s_k = n \right. \right\}$, then

$$\text{Invariant } \mathcal{R}_n = \{ \mathbf{u} \in \mathbb{R}^N | \mathbf{u} = a\mathbf{1} \text{ for all } a \in \mathbb{R} \},$$

$$\overrightarrow{\mathcal{R}_n} = \left\{ \mathbf{u} \in \mathbb{R}^N \left| \sum_{k \in U} x_k = 0 \right. \right\},$$

$$\boldsymbol{\lambda}_b = \mathbf{1} \frac{1}{N} \sum_{k \in U} \lambda_k = \mathbf{1}\bar{\lambda},$$

and

$$\boldsymbol{\lambda}_a = (\lambda_1 - \bar{\lambda} \quad \cdots \quad \lambda_k - \bar{\lambda} \quad \cdots \quad \lambda_N - \bar{\lambda}),$$

where $\mathbf{1}$ is a vector of ones of $\mathbb{R}^N$.

The main result of the theory of exponential families establishes that the mapping between the parameter $\boldsymbol{\lambda}$ and the expectation $\boldsymbol{\mu}$ is bijective.

Theorem 5. *Let* $p_{\mathrm{EXP}}(\mathbf{s}, \boldsymbol{\lambda}, \mathcal{Q})$ *be an exponential design on support* $\mathcal{Q}$. *Then*

$$\boldsymbol{\mu}(\boldsymbol{\lambda}) = \sum_{\mathbf{s} \in \mathcal{Q}} \mathbf{s} p_{\mathrm{EXP}}(\mathbf{s}, \boldsymbol{\lambda}, \mathcal{Q})$$

is a homeomorphism of $\overrightarrow{\mathcal{Q}}$ *(the direction of* $\mathcal{Q}$*) and* $\overset{\circ}{\mathcal{Q}}$ *(the interior of* $\mathcal{Q}$*); that is,* $\boldsymbol{\mu}(\boldsymbol{\lambda})$ *is a continuous and bijective function from* $\overrightarrow{\mathcal{Q}}$ *to* $\overset{\circ}{\mathcal{Q}}$ *whose inverse is also continuous.*

Theorem 5 is an application to exponential designs of a well-known result of exponential families. The proof is given, for instance, in Brown (1986, p. 74).

Example 9. If the support is $\mathcal{R}_n$, then $\boldsymbol{\mu}(\boldsymbol{\lambda})$ is bijective from

$$\overrightarrow{\mathcal{R}_n} = \left\{ \mathbf{u} \in \mathbb{R}^N \;\middle|\; \sum_{k \in U} u_k = 0 \right\},$$

to

$$\overset{\circ}{\mathcal{R}_n} = \left\{ \mathbf{u} \in]0, n[^N \;\middle|\; \sum_{k \in U} u_k = n \right\}.$$

Another important property of the exponential design is that the parameter does not change while conditioning with respect to a subset of the support.

Result 19. *Let* $\mathcal{Q}_1$ *and* $\mathcal{Q}_2$ *be two supports such that* $\mathcal{Q}_2 \subset \mathcal{Q}_1$. *Also, let*

$$p_1(\mathbf{s}) = g(\mathbf{s}) \exp\left[\boldsymbol{\lambda}'\mathbf{s} - \alpha(\boldsymbol{\lambda}, \mathcal{Q}_1)\right]$$

be an exponential design with support $\mathcal{Q}_1$. *Then*

$$p_2(\mathbf{s}) = p_1(\mathbf{s}|\mathcal{Q}_2)$$

is also an exponential sampling design with the same parameter.

The proof is obvious.

5.3 Poisson Sampling Design With Replacement (POISSWR)

5.3.1 Sampling Design

Definition 46. *An exponential design defined on* $\mathcal{R}$ *is called a Poisson sampling with replacement (POISSWR).*

The Poisson sampling with replacement can be derived from Definition 45, page 45.

$$p_{\mathrm{POISSWR}}(\mathbf{s},\boldsymbol{\lambda}) = p_{\mathrm{EXP}}(\mathbf{s},\boldsymbol{\lambda},\mathcal{R}) = \left(\prod_{k\in U}\frac{1}{s_k!}\right)\exp\left[\boldsymbol{\lambda}'\mathbf{s} - \alpha(\boldsymbol{\lambda},\mathcal{R})\right],$$

for all $\mathbf{s}\in\mathcal{R}$, and with $\boldsymbol{\lambda}\in\mathbb{R}^N$.

The normalizing constant is

$$\alpha(\boldsymbol{\lambda},\mathcal{R}) = \log\sum_{\mathbf{s}\in\mathcal{R}}\frac{\exp\boldsymbol{\lambda}'\mathbf{s}}{\prod_{k\in U}s_k!} = \log\sum_{\mathbf{s}\in\mathcal{R}}\prod_{k\in U}\frac{(\exp\lambda_k)^{s_k}}{s_k!}$$

$$= \log\prod_{k\in U}\sum_{j=1}^{\infty}\frac{(\exp\lambda_k)^j}{j!} = \log\prod_{k\in U}\exp(\exp\lambda_k) = \sum_{k\in U}\exp\lambda_k.$$

Thus,

$$\mu_k = \frac{\partial\sum_{k\in U}\exp\lambda_k}{\partial\lambda_k} = \exp\lambda_k,$$

$$\Sigma_{kk} = \frac{\partial^2\sum_{k\in U}\exp\lambda_k}{\partial\lambda_k^2} = \exp\lambda_k = \mu_k,$$

$$\Sigma_{k\ell} = \frac{\partial^2\sum_{k\in U}\exp\lambda_k^2}{\partial\lambda_k\partial\lambda_\ell} = 0, \quad\text{with } k\neq\ell,$$

and

$$\boldsymbol{\Sigma} = \mathrm{diag}(\mu_1\ \cdots\ \mu_k\ \cdots\ \mu_N),$$

which allows reformulating the definition of Poisson sampling with replacement.

Definition 47. *A sampling design* $p_{\mathrm{POISSWR}}(.,\boldsymbol{\lambda})$ *on* $\mathcal{R}$ *is said to be a Poisson sampling with replacement if it can be written*

$$p_{\mathrm{POISSWR}}(\mathbf{s},\boldsymbol{\lambda}) = \prod_{k\in U}\frac{\mu_k^{s_k}e^{-\mu_k}}{s_k!},$$

for all $\mathbf{s}\in\mathcal{R}$, *and with* $\mu_k\in\mathbb{R}_+^*$.

The S_k's thus have independent Poisson distributions: $S_k\sim\mathcal{P}(\mu_k)$. If the μ_k's are given, then $\boldsymbol{\lambda}$ can be computed by $\lambda_k = \log\mu_k, k\in U$. The characteristic function is

$$\phi_{\mathrm{POISSWR}}(\mathbf{t}) = \exp\left[\alpha(i\mathbf{t}+\boldsymbol{\lambda},\mathcal{Q}) - \alpha(\boldsymbol{\lambda},\mathcal{Q})\right]$$

$$= \exp\left[\sum_{k\in U}\exp(it_k+\lambda_k) - \sum_{k\in U}\exp\lambda_k\right] = \exp\sum_{k\in U}\mu_k(\exp it_k - 1).$$

5.3.2 Estimation

The Hansen-Hurwitz estimator is

$$\widehat{Y}_{HH} = \sum_{k \in U} \frac{y_k S_k}{\mu_k}.$$

Its variance is

$$\mathrm{var}(\widehat{Y}_{HH}) = \sum_{k \in U} \frac{y_k^2}{\mu_k}$$

and can be estimated by

$$\widehat{\mathrm{var}}_1(\widehat{Y}_{HH}) = \sum_{k \in U} S_k \frac{y_k^2}{\mu_k^2}.$$

However, because

$$\mathrm{E}\left[S_k | r(\mathbf{S})\right] = \begin{cases} \dfrac{\mu_k}{1 - e^{-\mu_k}} & \text{if } S_k > 0 \\ 0 & \text{if } S_k = 0, \end{cases}$$

the improved Hansen-Hurwitz estimator can be computed

$$\widehat{Y}_{IHH} = \sum_{k \in U} \frac{y_k \mathrm{E}[S_k | r(\mathbf{S})]}{\mu_k} = \sum_{k \in U} \frac{y_k r(S_k)}{1 - e^{-\mu_k}}.$$

Because

$$\mathrm{E}[r(S_k)] = 1 - \mathrm{Pr}(S_k = 0) = 1 - e^{-\mu_k},$$

and

$$\mathrm{var}[r(S_k)] = \mathrm{E}[r(S_k)]\left\{1 - \mathrm{E}[r(S_k)]\right\} = e^{-\mu_k}\left(1 - e^{-\mu_k}\right),$$

we obtain

$$\mathrm{var}(\widehat{Y}_{IHH}) = \sum_{k \in U} e^{-\mu_k}\left(1 - e^{-\mu_k}\right) \frac{y_k^2}{(1 - e^{-\mu_k})^2} = \sum_{k \in U} \frac{y_k^2}{e^{\mu_k} - 1},$$

and

$$\widehat{\mathrm{var}}(\widehat{Y}_{IHH}) = \sum_{k \in U} \frac{y_k^2}{(e^{\mu_k} - 1)(1 - e^{-\mu_k})}.$$

The improvement brings an important decrease of the variance with respect to the Hansen-Hurwitz estimator.

Finally, the Horvitz-Thompson estimator is

$$\widehat{Y}_{HT} = \sum_{k \in U} \frac{y_k r(S_k)}{\mathrm{E}[r(S_k)]} = \sum_{k \in U} \frac{y_k r(S_k)}{1 - e^{-\mu_k}}$$

and is equal to the improved Hansen-Hurwitz estimator.

Algorithm 5.1 Sequential procedure for POISSWR

DEFINITION k : INTEGER;
FOR $k = 1, \ldots, N$ DO
    randomly select s_k times unit k according to the Poisson distribution $\mathcal{P}(\mu_k)$;
ENDFOR.

5.3.3 Sequential Procedure for POISSWR

The standard sequential procedure (see Algorithm 3.2, page 34) applied to POISSWR gives the strictly sequential Algorithm 5.1.

5.4 Multinomial Design (MULTI)

5.4.1 Sampling Design

Definition 48. *An exponential design defined on $\mathcal{R}_n$ is called a multinomial design.*

The multinomial design can be derived from Definition 45, page 45.

$$p_{\mathrm{MULTI}}(\mathbf{s}, \boldsymbol{\lambda}, n) = p_{\mathrm{EXP}}(\mathbf{s}, \boldsymbol{\lambda}, \mathcal{R}_n) = \left(\prod_{k \in U} \frac{1}{s_k!} \right) \exp \left[\boldsymbol{\lambda}'\mathbf{s} - \alpha(\boldsymbol{\lambda}, \mathcal{R}_n) \right],$$

for all $\mathbf{s} \in \mathcal{R}_n$, and with $\boldsymbol{\lambda} \in \mathbb{R}^N$. The normalizing constant is

$$\alpha(\boldsymbol{\lambda}, \mathcal{R}_n) = \log \sum_{\mathbf{s} \in \mathcal{R}_n} \frac{\exp \boldsymbol{\lambda}'\mathbf{s}}{\prod_{k \in U} s_k!}$$

$$= \log \sum_{\mathbf{s} \in \mathcal{R}_n} \prod_{k \in U} \frac{(\exp \lambda_k)^{s_k}}{s_k!} = \log \frac{\left(\sum_{k \in U} \exp \lambda_k \right)^n}{n!}.$$

Thus,

$$\mu_k = \frac{\partial \alpha(\boldsymbol{\lambda}, \mathcal{R}_n)}{\partial \lambda_k} = \frac{n \exp \lambda_k}{\sum_{\ell \in U} \exp \lambda_\ell},$$

and

$$\Sigma_{k\ell} = \begin{cases} \dfrac{\partial^2 \alpha(\boldsymbol{\lambda}, \mathcal{R}_n)}{\partial \lambda_k^2} = \dfrac{\mu_k(n - \mu_k)}{n}, & k = \ell \\[4mm] \dfrac{\partial^2 \alpha(\boldsymbol{\lambda}, \mathcal{R}_n)}{\partial \lambda_k \partial \lambda_\ell} = -\dfrac{n \exp \lambda_k \exp \lambda_\ell}{\left(\sum_{j \in U} \exp \lambda_j \right)^2} = -\dfrac{\mu_k \mu_\ell}{n}, & k \neq \ell. \end{cases}$$

The variance-covariance operator is thus

$$\boldsymbol{\Sigma} = \mathrm{var}(\mathbf{S}) = \begin{pmatrix} \frac{\mu_1(n-\mu_1)}{n} & \cdots & \frac{-\mu_1\mu_k}{n} & \cdots & \frac{-\mu_1\mu_N}{n} \\ \vdots & \ddots & \vdots & & \vdots \\ \frac{-\mu_1\mu_k}{n} & \cdots & \frac{\mu_k(n-\mu_k)}{n} & \cdots & \frac{-\mu_k\mu_N}{n} \\ \vdots & & \vdots & \ddots & \vdots \\ \frac{-\mu_1\mu_N}{n} & \cdots & \frac{-\mu_k\mu_N}{n} & \cdots & \frac{\mu_N(n-\mu_N)}{n} \end{pmatrix}.$$

Moreover, we have

$$\mu_{k\ell} = \begin{cases} \dfrac{\mu_k(n-\mu_k)}{n} + \mu_k^2 = \mu_k + \dfrac{\mu_k^2(n-1)}{n}, & k = \ell \\[2mm] -\dfrac{\mu_k\mu_\ell}{n} + \mu_k\mu_\ell = \dfrac{\mu_k\mu_\ell(n-1)}{n}, & k \neq \ell, \end{cases}$$

and thus

$$\frac{\Sigma_{k\ell}}{\mu_{k\ell}} = \begin{cases} \dfrac{(n-\mu_k)}{n+\mu_k(n-1)}, & k = \ell \\[2mm] -\dfrac{1}{n-1}, & k \neq \ell. \end{cases} \tag{5.4}$$

The computation of $\boldsymbol{\mu}$ allows reformulating the multinomial design in function of the μ_k's.

Definition 49. *A sampling design $p_{\mathrm{MULTI}}(., \boldsymbol{\lambda}, n)$ on $\mathcal{R}_n$ is said to be a multinomial design (MULTI) if it can be written*

$$p_{\mathrm{MULTI}}(\mathbf{s}, \boldsymbol{\lambda}, n) = \frac{n!}{s_1! \dots s_k! \dots s_N!} \prod_{k \in U} (\mu_k/n)^{s_k},$$

for all $\mathbf{s} \in \mathcal{R}_n$, and where $\mu_k \in \mathbb{R}_+^*$ and

$$\sum_{k \in U} \mu_k = n.$$

This new formulation clearly shows that $\mathbf{S}$ has a multinomial distribution. The S_k's thus have non-independent binomial distributions: $S_k \sim \mathcal{B}(n, \mu_k/n)$.

The invariant subspace spanned by support $\mathcal{R}_n$ is $c \times \mathbf{1}$, for any value of $c \in \mathbb{R}$. If the μ_k are fixed then $\boldsymbol{\lambda}$ can be computed by $\lambda_k = \log \mu_k + c, k \in U$, for any value of $c \in \mathbb{R}$. The characteristic function is

$$\phi_{\mathrm{MULTI}}(\mathbf{t}) = \left(\frac{1}{n} \sum_{k \in U} \mu_k \exp it_k \right)^n.$$

5.4.2 Estimation

1. The Hansen-Hurwitz estimator is

$$\widehat{Y}_{HH} = \sum_{k \in U} S_k \frac{y_k}{\mu_k},$$

and its variance is

$$\text{var}(\widehat{Y}_{HH}) = \sum_{k \in U} \frac{y_k^2}{\mu_k} - \frac{1}{n} \left(\sum_{k \in U} y_k \right)^2 = \sum_{k \in U} \frac{\mu_k}{n^2} \left(\frac{n y_k}{\mu_k} - Y \right)^2.$$

2. With Expression (5.4), the estimator (2.19), page 28, becomes

$$\widehat{\text{var}}_2(\widehat{Y}_{HH}) = \frac{n}{(n-1)} \sum_{k \in U} S_k \left(\frac{y_k}{\mu_k} - \frac{\widehat{Y}_{HH}}{n} \right)^2. \tag{5.5}$$

Using Result 9, page 27, estimator (2.18), page 28, becomes

$$\widehat{\text{var}}_1(\widehat{Y}_{HH}) = \widehat{\text{var}}_2(\widehat{Y}_{HH}) + \sum_{k \in U} \frac{S_k y_k^2}{\mu_k^2} \sum_{\ell \in U} \frac{S_\ell \Sigma_{k\ell}}{\mu_{k\ell}}$$

$$= \frac{n}{(n-1)} \sum_{k \in U} S_k \left(\frac{y_k}{\mu_k} - \frac{\widehat{Y}_{HH}}{n} \right)^2$$

$$+ \sum_{k \in U} \frac{y_k^2}{\mu_k^2} \frac{n}{n-1} \left\{ S_k^2 \frac{n}{[n + \mu_k(n-1)]} - S_k \right\}. \tag{5.6}$$

As for SRSWR, this estimator (5.6) should never be used.

3. Because

$$\pi_k = \Pr(S_k > 0) = 1 - \Pr(S_k = 0) = 1 - \left(1 - \frac{\mu_k}{n} \right)^n,$$

and

$$\pi_{k\ell} = \Pr(S_k > 0 \text{ and } S_\ell > 0) = 1 - \Pr(S_k = 0 \text{ or } S_\ell = 0)$$

$$= 1 - \Pr(S_k = 0) - P(S_\ell = 0) + \Pr(S_k = 0 \text{ and } S_\ell = 0)$$

$$= 1 - \left[1 - \left(1 - \frac{\mu_k}{n} \right)^n \right] - \left[1 - \left(1 - \frac{\mu_\ell}{n} \right)^n \right] + 1 - \left(1 - \frac{\mu_k}{n} - \frac{\mu_k}{n} \right)^n$$

$$= \left(1 - \frac{\mu_k}{n} \right)^n + \left(1 - \frac{\mu_\ell}{n} \right)^n - \left(1 - \frac{\mu_k}{n} - \frac{\mu_k}{n} \right)^n,$$

the Horvitz-Thompson estimator is

$$\widehat{Y}_{HT} = \sum_{k \in U} \frac{r(S_k) y_k}{1 - \left(1 - \frac{\mu_k}{n} \right)^n}.$$

Its variance and variance estimator can be constructed by means of the joint inclusion probabilities.

4. The construction of the improved Hansen-Hurwitz estimator seems to be an unsolved problem. The main difficulty consists of calculating $\mathrm{E}[S_k|r(\mathbf{S})]$. A solution can be given for $n = 3$ but seems to be complicated for larger sample sizes.

Case where $n(\mathbf{S}) = 3$

Let $\mathbf{a}_k = (0 \cdots 0 \underbrace{1}_{k\text{th}} 0 \cdots 1)' \in \mathbb{R}^N$. Defining $q_k = \mu_k/n$, we have

$$\Pr[r(\mathbf{S}) = \mathbf{a}_k] = \Pr[\mathbf{S} = 3\mathbf{a}_k] = q_k^3, \quad k \in U,$$

$$\Pr[r(\mathbf{S}) = \mathbf{a}_k + \mathbf{a}_\ell] = \Pr(\mathbf{S} = 2\mathbf{a}_k + \mathbf{a}_\ell) + \Pr(\mathbf{S} = \mathbf{a}_k + 2\mathbf{a}_\ell)$$
$$= 3q_k^2 q_\ell + 3q_k q_\ell^2 = 3q_k q_\ell(q_k + q_\ell), \quad q \neq \ell \in U,$$

$$\Pr[r(\mathbf{S}) = \mathbf{a}_k + \mathbf{a}_\ell + \mathbf{a}_m] = \Pr(\mathbf{S} = \mathbf{a}_k + \mathbf{a}_\ell + \mathbf{a}_m)$$
$$= \frac{3!}{1!1!1!} q_k q_\ell q_m = 6q_k q_\ell q_m.$$

The distribution of $n[r(s)]$ can thus be computed

$$\Pr\{n[r(\mathbf{S})] = 1\} = \sum_{\ell \in U} q_k^3, \quad k \in U,$$

$$\Pr\{n[r(\mathbf{S})] = 2\} = \sum_{k \in U} \sum_{\substack{\ell \in U \\ \ell < k}} 3q_k q_\ell(q_k + q_\ell)$$
$$= 3 \sum_{k \in U} q_k^2 - 3 \sum_{k \in U} q_k^3, \quad k \neq \ell \in U,$$

$$\Pr\{n[r(\mathbf{S})] = 3\} = \sum_{k \in U} \sum_{\substack{\ell \in U \\ \ell < k}} \sum_{\substack{m \in U \\ m < \ell}} 6q_k q_\ell q_m$$
$$= 1 - 3 \sum_{k \in U} q_k^2 + 2 \sum_{k \in U} q_k^3, \quad k \neq \ell \neq m \in U.$$

The first two conditional expectations are simple:

$$\mathrm{E}[\mathbf{S}|r(\mathbf{S}) = \mathbf{a}_k] = 3\mathbf{a}_k, \quad \text{for all } k \in U,$$

$$\mathrm{E}[\mathbf{S}|r(\mathbf{S}) = \mathbf{a}_k + \mathbf{a}_\ell + \mathbf{a}_m] = \mathbf{a}_k + \mathbf{a}_\ell + \mathbf{a}_m, \quad \text{for all } k \neq \ell \neq m \in U.$$

In the case where $n[r(\mathbf{S})] = 2$, we have

$$
\begin{aligned}
&\mathrm{E}[S_k|r(\mathbf{S}) = \mathbf{a}_k + \mathbf{a}_\ell]\\
&= 1 \times \Pr(S_k = 1|r(\mathbf{S}) = \mathbf{a}_k + \mathbf{a}_\ell) + 2 \times \Pr(S_k = 1|r(\mathbf{S}) = \mathbf{a}_k + \mathbf{a}_\ell)\\
&= 1 \times \frac{\Pr(S_k = 1 \text{ and } r(\mathbf{S}) = \mathbf{a}_k + \mathbf{a}_\ell)}{\Pr(r(\mathbf{S}) = \mathbf{a}_k + \mathbf{a}_\ell)}\\
&\quad + 2 \times \frac{\Pr(S_k = 2 \text{ and } r(\mathbf{S}) = \mathbf{a}_k + \mathbf{a}_\ell)}{\Pr(r(\mathbf{S}) = \mathbf{a}_k + \mathbf{a}_\ell)}\\
&= 1 \times \frac{\Pr(S_k = 1 \text{ and } S_\ell = 2)}{\Pr(r(\mathbf{S}) = \mathbf{a}_k + \mathbf{a}_\ell)} + 2 \times \frac{\Pr(S_k = 2 \text{ and } S_\ell = 1)}{\Pr(r(\mathbf{S}) = \mathbf{a}_k + \mathbf{a}_\ell)}\\
&= \frac{1 \times 3q_k q_\ell^2 + 2 \times 3q_k^2 q_\ell}{3q_k q_\ell^2 + 3q_k^2 q_\ell} = \frac{2q_k + q_\ell}{q_k + q_\ell}, \text{ for all } k \in U.
\end{aligned}
$$

The improved Hansen-Hurwitz estimator can be computed as follows

$$
\widehat{Y}_{IHH} = \sum_{k \in U} \frac{y_k \mathrm{E}[S_k|r(S)]}{\mu_k}.
$$

We have:

- If $r(\mathbf{S}) = \mathbf{a}_k$, then $\widehat{Y}_{IHH} = \dfrac{3y_k}{\mu_k}$.
- If $r(\mathbf{S}) = \mathbf{a}_k + \mathbf{a}_\ell$, then $\widehat{Y}_{IHH} = \dfrac{y_k}{\mu_k}\dfrac{2\mu_k + \mu_\ell}{\mu_k + \mu_\ell} + \dfrac{y_\ell}{\mu_\ell}\dfrac{\mu_k + 2\mu_\ell}{\mu_k + \mu_\ell}, k \neq \ell$.
- If $r(\mathbf{S}) = \mathbf{a}_k + \mathbf{a}_\ell + \mathbf{a}_m$, then $\widehat{Y}_{IHH} = \dfrac{y_k}{\mu_k} + \dfrac{y_\ell}{\mu_\ell} + \dfrac{y_m}{\mu_m}, k \neq \ell \neq m$.

Unfortunately, the generalization to $n > 3$ becomes very intricate.

5.4.3 Sequential Procedure for Multinomial Design

For the sequential implementation, consider the following result.

Result 20. *In multinomial design,*

$$
\begin{aligned}
q_k(s_k) &= \Pr(S_k = s_k|S_{k-1} = s_{k-1}, \ldots, S_1 = s_1)\\
&= \frac{\left(n - \sum_{\ell=1}^{k-1} s_\ell\right)!}{\left(n - \sum_{\ell=1}^{k} s_\ell\right)! s_k!} \left(\frac{\mu_k}{n - \sum_{\ell=1}^{k-1} \mu_\ell}\right)^{s_k} \left(1 - \frac{\mu_k}{n - \sum_{\ell=1}^{k-1} \mu_\ell}\right)^{n - \sum_{\ell=1}^{k} s_\ell}
\end{aligned}
$$

Proof. By definition

$$
\begin{aligned}
q_k(s_k) &= \Pr(S_k = s_k|S_{k-1} = s_{k-1}, \ldots, S_1 = s_1)\\
&= \frac{\sum_{\mathbf{s} \in \mathcal{R}_n | S_k = s_k, S_{k-1} = s_{k-1}, \ldots, S_1 = s_1} p_{\mathrm{EXP}}(\mathbf{s}, \boldsymbol{\lambda}, \mathcal{R}_n)}{\sum_{\mathbf{s} \in \mathcal{R}_n | S_{k-1} = s_{k-1}, \ldots, S_1 = s_1} p_{\mathrm{EXP}}(\mathbf{s}, \boldsymbol{\lambda}, \mathcal{R}_n)}.
\end{aligned}
$$

Moreover,

$$\sum_{\substack{\mathbf{s}\in\mathcal{R}_n \\ S_k=s_k, S_{k-1}=s_{k-1},\ldots,S_1=s_1}} p_{\text{EXP}}(\mathbf{s}, \boldsymbol{\lambda}, \mathcal{R}_n)$$

$$= \sum_{\substack{\mathbf{s}\in\mathcal{R}_n \\ S_k=s_k, S_{k-1}=s_{k-1},\ldots,S_1=s_1}} n! \prod_{\ell=1}^{N} \frac{(\mu_\ell/n)^{s_\ell}}{s_\ell!}$$

$$= \frac{n!}{\left(\sum_{\ell=k+1}^{N} s_\ell\right)!} \prod_{\ell=1}^{k} \frac{(\mu_\ell/n)^{s_\ell}}{s_\ell!} \sum_{\substack{\mathbf{s}\in\mathcal{R}_n \\ S_k=s_k, S_{k-1}=s_{k-1},\ldots,S_1=s_1}} \left(\sum_{\ell=k+1}^{N} s_\ell\right)! \prod_{\ell=k+1}^{N} \frac{(\mu_\ell/n)^{s_\ell}}{s_\ell!}$$

$$= \frac{n!}{\left(\sum_{\ell=k+1}^{N} s_\ell\right)!} \prod_{\ell=1}^{k} \frac{(\mu_\ell/n)^{s_\ell}}{s_\ell!} \left(\sum_{\ell=k+1}^{N} \frac{\mu_\ell}{n}\right)^{\sum_{\ell=k+1}^{N} s_\ell}.$$

By the same reasoning, we get

$$\sum_{\substack{\mathbf{s}\in\mathcal{R}_n \\ S_{k-1}=s_{k-1},\ldots,S_1=s_1}} p_{\text{EXP}}(\mathbf{s}, \boldsymbol{\lambda}, \mathcal{R}_n) = \frac{n!}{\left(\sum_{\ell=k}^{N} s_\ell\right)!} \prod_{\ell=1}^{k-1} \frac{(\mu_\ell/n)^{s_\ell}}{s_\ell!} \left(\sum_{\ell=k}^{N} \frac{\mu_\ell}{n}\right)^{\sum_{\ell=k}^{N} s_\ell}.$$

We thus have

$$q_k(s_k) = \frac{\left(\sum_{\ell=k}^{N} s_\ell\right)!}{\left(\sum_{\ell=k+1}^{N} s_\ell\right)! s_k!} \left(\frac{\mu_k/n}{\sum_{\ell=k}^{N} \mu_\ell/n}\right)^{s_k} \left(\frac{\sum_{\ell=k+1}^{N} \mu_\ell/n}{\sum_{\ell=k}^{N} \mu_\ell/n}\right)^{\sum_{\ell=k+1}^{N} s_\ell}$$

$$= \frac{\left(n-\sum_{\ell=1}^{k-1} s_\ell\right)!}{\left(n-\sum_{\ell=1}^{k} s_\ell\right)! s_k!} \left(\frac{\mu_k}{n-\sum_{\ell=1}^{k-1} \mu_\ell}\right)^{s_k} \left(1-\frac{\mu_k}{n-\sum_{\ell=1}^{k-1} \mu_\ell}\right)^{n-\sum_{\ell=1}^{k} s_\ell}. \qquad \square$$

By the standard sequential Algorithm 3.2, page 34 and Result 20, we can define the very simple strictly sequential procedure in Algorithm 5.2.

Algorithm 5.2 Sequential procedure for multinomial design

DEFINITION k : INTEGER;
FOR $k = 1, \ldots, N$ DO
    select the kth unit s_k times according to the binomial distribution

$$\mathcal{B}\left(n - \sum_{\ell=1}^{k-1} s_\ell, \frac{\mu_k}{n - \sum_{\ell=1}^{k-1} \mu_\ell}\right);$$

ENDFOR.

This procedure was first proposed by Kemp and Kemp (1987) in another context as survey sampling (see also Bol'shev, 1965; Brown and Bromberg,

1984; Dagpunar, 1988; Devroye, 1986; Davis, 1993; Loukas and Kemp, 1983; Johnson et al., 1997, pp. 67-69).

5.4.4 Draw by Draw Procedure for Multinomial Design

The standard draw by draw Algorithm 3.3, page 35, gives the well-known ball-in-urn method (on this topic see Hansen and Hurwitz, 1943; Dagpunar, 1988; Devroye, 1986; Ho et al., 1979; Johnson et al., 1997, p. 68) presented in Algorithm 5.3.

Algorithm 5.3 Draw by draw procedure for multinomial design

DEFINITION j : INTEGER;
FOR $j = 1, \ldots, n$ DO
  a unit is selected with probability μ_k/n from the population U;
ENDFOR.

5.5 Poisson Sampling Without Replacement (POISSWOR)

5.5.1 Sampling Design

Poisson sampling design was studied by Hájek (1964), Ogus and Clark (1971), Brewer et al. (1972), Brewer et al. (1984), and Cassel et al. (1993, p. 17).

Definition 50. *An exponential design defined on $\mathcal{S}$ is called a Poisson sampling design.*

The Poisson design can be derived from Definition 45, page 45.

$$p_{\text{POISSWOR}}(\mathbf{s}, \boldsymbol{\lambda}) = p_{\text{EXP}}(\mathbf{s}, \boldsymbol{\lambda}, \mathcal{S}) = \exp\left[\boldsymbol{\lambda}'\mathbf{s} - \alpha(\boldsymbol{\lambda}, \mathcal{S})\right],$$

for all $\mathbf{s} \in \mathcal{S}$, and with $\boldsymbol{\lambda} \in \mathbb{R}^N$. A simplification is derived from the fact that $s_k! = 1$ because s_k takes only the values 1 or 0. The normalizing constant is

$$\alpha(\boldsymbol{\lambda}, \mathcal{S}) = \log \sum_{\mathbf{s} \in \mathcal{S}} \exp \boldsymbol{\lambda}'\mathbf{s} = \log \prod_{k \in U} (1 + \exp \lambda_k). \tag{5.7}$$

Thus,

$$\pi_k = \frac{\partial \alpha(\boldsymbol{\lambda}, \mathcal{S})}{\partial \lambda_k} = \frac{\exp \lambda_k}{1 + \exp \lambda_k}, \tag{5.8}$$

$$\Delta_{k\ell} = \begin{cases} \dfrac{\partial^2 \alpha(\boldsymbol{\lambda}, \mathcal{S})}{\partial \lambda_k^2} = \dfrac{\exp \lambda_k}{1 + \exp \lambda_k} \left(1 - \dfrac{\exp \lambda_k}{1 + \exp \lambda_k}\right) = \pi_k(1 - \pi_k), & k = \ell \\[4mm] \dfrac{\partial^2 \alpha(\boldsymbol{\lambda}, \mathcal{S})}{\partial \lambda_k \partial \lambda_\ell} = 0, & k \neq \ell, \end{cases}$$

and $\pi_{k\ell} = \pi_k \pi_\ell, k \neq \ell$.

The variance-covariance operator is

$$\boldsymbol{\Delta} = \text{diag}\left[\pi_1(1 - \pi_1) \cdots \pi_k(1 - \pi_k) \cdots \pi_N(1 - \pi_N)\right].$$

The computation of $\boldsymbol{\pi}$ allows reformulating the sampling design as a function of π_k.

Definition 51. *A sampling design* $p_{\text{POISSWOR}}(., \boldsymbol{\lambda})$ *on* $\mathcal{S}$ *is said to be a Poisson sampling without replacement if it can be written*

$$p_{\text{POISSWOR}}(\mathbf{s}, \boldsymbol{\lambda}) = \prod_{k \in U} \left[\pi_k^{s_k}(1 - \pi_k)^{1 - s_k}\right],$$

for all $\mathbf{s} \in \mathcal{S}$.

The vector $\boldsymbol{\lambda}$ can be computed easily from the π_k by

$$\lambda_k = \log \frac{\pi_k}{1 - \pi_k}.$$

The characteristic function is

$$\phi_{\text{POISSWOR}}(\mathbf{t}) = \prod_{k \in U} \left[1 + \pi_k \left(\exp it_k - 1\right)\right].$$

5.5.2 Distribution of $n(\mathbf{S})$

In Poisson sampling, the random variable $n(\mathbf{S})$ has the possible values $0, 1, \ldots, N$, and is a sum of independent Bernoulli trials, which is usually called a Poisson-binomial distribution (see among others Hodges and LeCam, 1960; Chen and Liu, 1997).

Chen and Liu (1997) and Aires (1999) have pointed out that it is possible to compute recursively the distribution of $n(\mathbf{S})$. This result was, however, already mentioned in Sampford (1967, p. 507) in another context. Let

$$U_i^j = \{i, i+1, \ldots, j-1, j\} \quad \text{with } 1 \leq i \leq j, j \leq N,$$

$$\mathcal{S}_z(U_i^j) = \left\{\mathbf{s} \in \mathcal{S}_z | s_k = 0, \text{ for } k \notin U_i^j\right\}, \quad \text{with } z \leq i - j + 1,$$

$$G\left[\boldsymbol{\lambda}, \mathcal{S}_z(U_i^j)\right] = \sum_{\mathbf{s} \in \mathcal{S}_z(U_i^j)} p_{\text{POISSWOR}}(\mathbf{s}, \boldsymbol{\lambda}) = \sum_{\mathbf{s} \in \mathcal{S}_z(U_i^j)} \prod_{k=i}^{j} \pi_k^{s_k}(1 - \pi_k)^{1 - s_k},$$

and

$$\pi_k = \frac{\exp \lambda_k}{1 + \exp \lambda_k}, \quad k \in U.$$

We have

$$\Pr[n(\mathbf{S}) = z] = \frac{\sum_{s \in \mathcal{S}_z} \prod_{k=1}^N \pi_k^{s_k}(1 - \pi_k)^{1-s_k}}{\sum_{s \in \mathcal{S}} \prod_{k=1}^N \pi_k^{s_k}(1 - \pi_k)^{1-s_k}}$$

$$= \frac{\exp \alpha(\boldsymbol{\lambda}, \mathcal{S}_z)}{\exp \alpha(\boldsymbol{\lambda}, \mathcal{S})} = G\left[\boldsymbol{\lambda}, \mathcal{S}_z(U_1^N)\right].$$

This probability can be computed recursively by means of the following relation

$$G\left[\boldsymbol{\lambda}, \mathcal{S}_z(U_1^j)\right] = G\left[\boldsymbol{\lambda}, \mathcal{S}_{z-1}(U_1^{j-1})\right]\pi_j + G\left[\boldsymbol{\lambda}, \mathcal{S}_z(U_1^{j-1})\right](1 - \pi_j), \quad (5.9)$$

for $j = z, \ldots, N$. An example is given in Table 5.1. Equivalently,

$$G\left[\boldsymbol{\lambda}, \mathcal{S}_z(U_i^N)\right] = G\left[\boldsymbol{\lambda}, \mathcal{S}_{z-1}(U_{i+1}^N)\right]\pi_i + G\left[\boldsymbol{\lambda}, \mathcal{S}_z(U_{i+1}^N)\right](1 - \pi_i), \quad (5.10)$$

for $i = 1, \ldots, N - z$.

Table 5.1. Example of a recursive construction of $G\left[\boldsymbol{\lambda}, \mathcal{S}_z(U_1^j)\right]$ for $N = 10$

π_k		$j = 0$	1	2	3	4	5	6	Total
	$z = 0$	1							1
0.10208	1	0.89792	0.10208						1
0.22382	2	0.69695	0.28020	0.02285					1
0.44173	3	0.38909	0.46429	0.13653	0.01009				1
0.57963	4	0.16356	0.42070	0.32651	0.08338	0.00585			1
0.77935	5	0.03609	0.22030	0.39992	0.27286	0.06627	0.00456		1
0.87340	6	0.00457	0.05941	0.24304	0.38383	0.24671	0.05846	0.00398	1

5.5.3 Estimation

The Horvitz-Thompson estimator is

$$\widehat{Y}_{HT} = \sum_{k \in U} \frac{y_k S_k}{\pi_k}.$$

Its variance is

$$\mathrm{var}(\widehat{Y}_{HT}) = \sum_{k \in U} \pi_k(1 - \pi_k)\frac{y_k^2}{\pi_k^2},$$

and its unbiased estimator

$$\widehat{\mathrm{var}}(\widehat{Y}_{HT}) = \sum_{k \in U}(1 - \pi_k)\frac{y_k^2}{\pi_k^2}.$$

5.5.4 Sequential Procedure for POISSWOR

The implementation of Algorithm 3.2, page 34, to POISSWOR is presented in the strictly sequential Algorithm 5.4.

Algorithm 5.4 Sequential procedure for POISSWOR

DEFINITION k : INTEGER;
FOR $k = 1, \ldots, N$, DO select the kth unit with probability π_k;
ENDFOR.

5.6 Conditional Poisson Sampling (CPS)

5.6.1 Sampling Design

The most interesting exponential design is defined with support $\mathcal{S}_n$, but its implementation is more intricate. It is generally called Conditional Poisson Sampling (CPS) because it can be obtained by selecting samples by means of a Poisson sampling design without replacement until a given sample size is obtained. However, this appellation is a bit unfortunate because conditioning a Poisson design with a fixed sample size is only a way to implement this design. Several authors refer to it as "exponential design without replacement" or "maximum entropy design" because it can be obtained by maximizing the entropy measure

$$I(p) = - \sum_{\mathbf{s} \in \mathcal{S}_n} p(\mathbf{s}) \log p(\mathbf{s}),$$

subject to given inclusion probabilities.

The CPS design can also be obtained by minimizing the Kullback-Leibler divergence sampling from a simple random sampling without replacement. It was studied by Hájek (1964) and is one of the principal topics of his posthumous book (Hájek, 1981). Hájek proposed to implement it by a Poisson rejective procedure, but the Hájek methods do not respect exactly the fixed inclusion probabilities (Brewer and Hanif, 1983, see procedure 28-31, pp. 40-41). The main problem with the implementation of this design is that the characteristic function cannot be simplified and that it seems impossible to compute the vector of inclusion probabilities without enumerating all the possible samples.

A very important result has, however, been given by Chen et al. (1994). They have proposed an algorithm that allows deriving the inclusion probabilities from the parameter and vice versa. In a manuscript paper, Deville (2000) has improved this algorithm. Chen et al. (1994), Chen and Liu (1997), Chen (1998, 2000), and Deville (2000) pointed out that a fast computation of the parameter allows a rapid implementation of this sampling design. Jonasson and Nerman (1996) and Aires (1999, 2000a) have independently investigated the Conditional Poisson Sampling. This recent progress allows constructing several procedures: rejective Poisson sampling, rejective multinomial design, sequential sampling, and the draw by draw procedure.

Definition 52. *An exponential design defined on $\mathcal{S}_n$ is called a Conditional Poisson Sampling (CPS) design.*

The CPS design can be derived from Definition 45, page 45.

$$p_{\mathrm{CPS}}(\mathbf{s}, \boldsymbol{\lambda}, n) = p_{\mathrm{EXP}}(\mathbf{s}, \boldsymbol{\lambda}, \mathcal{S}_n) = \exp\left[\boldsymbol{\lambda}'\mathbf{s} - \alpha(\boldsymbol{\lambda}, \mathcal{S}_n)\right],$$

for all $\mathbf{s} \in \mathcal{S}_n$, and with $\boldsymbol{\lambda} \in \mathbb{R}^N$.

When the support is $\mathcal{S}_n$, the problem becomes more intricate because $\alpha(\boldsymbol{\lambda}, \mathcal{S}_n)$ cannot be simplified. For this reason, one could have believed (before the paper of Chen et al., 1994) that it was not possible to select a sample with this design without enumerating all the samples of $\mathcal{S}_n$. We show that this sampling design can now be implemented quite easily.

5.6.2 Inclusion Probabilities

The vector of inclusion probabilities is defined by

$$\boldsymbol{\pi}(\boldsymbol{\lambda}, \mathcal{S}_n) = \sum_{\mathbf{s} \in \mathcal{S}_n} \mathbf{s} p_{\mathrm{EXP}}(\mathbf{s}, \boldsymbol{\lambda}, \mathcal{S}_n).$$

Because

$$\mathrm{Invariant}\,(\mathcal{S}_n) = \left\{\mathbf{x} \in \mathbb{R}^N | \mathbf{x} = a\mathbf{1}, \text{ for all } a \in \mathbb{R}\right\},$$

vector $\boldsymbol{\lambda}$ can be re-scaled in order that $\sum_{k\in U} \lambda_k = 0$, which provides a unique definition of $\boldsymbol{\lambda}$.

Because $\sum_{k\in U} \pi_k = n$, from Theorem 5, page 67, we obtain the particular Result 21 (see also Chen, 2000, Theorem 3.1).

Result 21. *The application $\boldsymbol{\pi}(\boldsymbol{\lambda}, \mathcal{S}_n)$ is bijective from*

$$\vec{\mathcal{S}_n} = \left\{\boldsymbol{\lambda} \in \mathbb{R}^N \left| \sum_{k\in U} \lambda_k = 0\right.\right\}$$

to

$$\overset{\circ}{\mathcal{S}_n} = \left\{\boldsymbol{\pi} \in\,]0, 1[^N \left| \sum_{k\in U} \pi_k = n\right.\right\}.$$

The first step consists of computing the inclusion probability $\boldsymbol{\pi}(\boldsymbol{\lambda}, \mathcal{S}_n)$ from $\boldsymbol{\lambda}$, which is theoretically given by

$$\boldsymbol{\pi}(\boldsymbol{\lambda}, \mathcal{S}_n) = \sum_{\mathbf{s} \in \mathcal{S}_n} \mathbf{s} p_{\mathrm{EXP}}(\mathbf{s}, \boldsymbol{\lambda}, \mathcal{S}_n) = \frac{\sum_{\mathbf{s}\in\mathcal{S}_n} \mathbf{s}\exp\boldsymbol{\lambda}'\mathbf{s}}{\sum_{\mathbf{s}\in\mathcal{S}_n} \exp\boldsymbol{\lambda}'\mathbf{s}}. \tag{5.11}$$

Unfortunately, Expression (5.11) becomes impossible to compute when U is large. The application of the CPS design seemed to be restricted only to very small populations before an important result due to Chen et al. (1994) and completed by Deville (2000). These authors have shown a recursive relation between $\pi(\boldsymbol{\lambda}, \mathcal{S}_{n-1})$ and $\pi(\boldsymbol{\lambda}, \mathcal{S}_n)$, which allows deriving $\pi(\boldsymbol{\lambda}, \mathcal{S}_n)$ from $\boldsymbol{\lambda}$ without enumerating all the possible samples of $\mathcal{S}$.

Result 22.

(i) $$\pi_k(\boldsymbol{\lambda}, \mathcal{S}_n) = (\exp \lambda_k)\left[1 - \pi_k(\boldsymbol{\lambda}, \mathcal{S}_{n-1})\right] \frac{\exp \alpha(\boldsymbol{\lambda}, \mathcal{S}_{n-1})}{\exp \alpha(\boldsymbol{\lambda}, \mathcal{S}_n)}, \tag{5.12}$$

(ii) $$\pi_k(\boldsymbol{\lambda}, \mathcal{S}_n) = n \frac{\exp \lambda_k \left[1 - \pi_k(\boldsymbol{\lambda}, \mathcal{S}_{n-1})\right]}{\sum_{\ell \in U} \exp \lambda_\ell \left[1 - \pi_\ell(\boldsymbol{\lambda}, \mathcal{S}_{n-1})\right]}, \tag{5.13}$$

(iii) $$\pi_k(\boldsymbol{\lambda}, \mathcal{S}_n) = \frac{\exp n\lambda_k}{\exp \alpha(\boldsymbol{\lambda}, \mathcal{S}_n)} \sum_{j=1}^{n} (-1)^{n-j} \frac{\exp \alpha(\boldsymbol{\lambda}, \mathcal{S}_{j-1})}{\exp(j-1)\lambda_k}. \tag{5.14}$$

Proof. (i) We have

$$\pi_k(\boldsymbol{\lambda}, \mathcal{S}_n) = \frac{\sum_{\mathbf{s} \in \mathcal{S}_n} s_k \exp \boldsymbol{\lambda}'\mathbf{s}}{\exp \alpha(\boldsymbol{\lambda}, \mathcal{S}_n)} = \frac{\exp \lambda_k}{\exp \alpha(\boldsymbol{\lambda}, \mathcal{S}_n)} \sum_{\substack{\mathbf{s} \in \mathcal{S}_{n-1} \\ s_k=0}} \exp \boldsymbol{\lambda}'\mathbf{s}$$

$$= \frac{\exp \lambda_k}{\exp \alpha(\boldsymbol{\lambda}, \mathcal{S}_n)} \left(\sum_{\mathbf{s} \in \mathcal{S}_{n-1}} \exp \boldsymbol{\lambda}'\mathbf{s} - \sum_{\mathbf{s} \in \mathcal{S}_{n-1}} s_k \exp \boldsymbol{\lambda}'\mathbf{s} \right)$$

$$= (\exp \lambda_k)\left[1 - \pi_k(\boldsymbol{\lambda}, \mathcal{S}_{n-1})\right] \frac{\exp \alpha(\boldsymbol{\lambda}, \mathcal{S}_{n-1})}{\exp \alpha(\boldsymbol{\lambda}, \mathcal{S}_n)}.$$

(ii) Because $\sum_{k \in U} \pi_k(\boldsymbol{\lambda}, \mathcal{S}_n) = n$, we get

$$n = \sum_{k \in U} \exp \lambda_k \left[1 - \pi_k(\boldsymbol{\lambda}, \mathcal{S}_{n-1})\right] \frac{\exp \alpha(\boldsymbol{\lambda}, \mathcal{S}_{n-1})}{\exp \alpha(\boldsymbol{\lambda}, \mathcal{S}_n)},$$

which gives Expression (5.13).

(iii) Expression (5.14) satisfies the recurrence equation (5.12).   □

Because $\pi_k(\boldsymbol{\lambda}, \mathcal{S}_0) = 0$, for all $k \in U$, several methods that implement the function $\pi(\boldsymbol{\lambda}, \mathcal{S}_n)$ can be used. If the $\alpha(\boldsymbol{\lambda}, \mathcal{S}_z), z = 1, \ldots, n$, are known, then Expression (5.14) can be used to compute $\pi(\boldsymbol{\lambda}, \mathcal{S}_n)$ directly. The recursive relation given in (5.13) allows defining a very quick algorithm to compute function $\pi(\boldsymbol{\lambda}, \mathcal{S}_n)$. An example is given in Table 5.2.

5.6.3 Computation of $\boldsymbol{\lambda}$ from Predetermined Inclusion Probabilities

The knowledge of $\boldsymbol{\lambda}$ allows us to compute the inclusion probabilities $\pi(\boldsymbol{\lambda}, \mathcal{S}_n)$ quickly. Nevertheless, the inclusion probabilities are generally fixed and the

Table 5.2. Recursive computation of $\pi(\boldsymbol{\lambda}, \mathcal{S}_n)$ from a given parameter $\boldsymbol{\lambda}$ by means of Expression (5.13)

k	λ_k	$\exp(\lambda_k)$	$n=0$	1	2	3	4	5	6
1	-2.151	0.116	0	0.009	0.028	0.07	0.164	0.401	1
2	-1.221	0.295	0	0.022	0.069	0.17	0.372	0.764	1
3	-0.211	0.810	0	0.061	0.182	0.41	0.726	0.914	1
4	0.344	1.411	0	0.106	0.301	0.61	0.837	0.951	1
5	1.285	3.614	0	0.272	0.629	0.83	0.934	0.981	1
6	1.954	7.059	0	0.531	0.792	0.91	0.966	0.990	1
	0	13.305	0	1	2	3	4	5	6

main problem is to compute $\boldsymbol{\lambda}$ from a given vector of inclusion probabilities $\boldsymbol{\pi}$. Suppose that $\widetilde{\boldsymbol{\pi}}$ is the vector of inclusion probabilities of a Poisson sampling design $\widetilde{p}_{\text{POISSWOR}}(.)$ with parameter equal to $\boldsymbol{\lambda} + c\mathbf{1}$; that is,

$$\widetilde{\boldsymbol{\pi}} = (\widetilde{\pi}_1 \cdots \widetilde{\pi}_k \cdots \widetilde{\pi}_N)' = \boldsymbol{\pi}(\boldsymbol{\lambda} + c\mathbf{1}, \mathcal{S}), c \in \mathbb{R}.$$

Note that by Result 18, page 66,

$$p_{\text{EXP}}(\mathbf{s}, \boldsymbol{\lambda}, \mathcal{S}_n) = \widetilde{p}_{\text{POISSWOR}}(\mathbf{s}, \boldsymbol{\lambda} + c\mathbf{1}|\mathcal{S}_n) = \frac{\widetilde{p}_{\text{POISSWOR}}(\mathbf{s}, \boldsymbol{\lambda} + c\mathbf{1})}{\sum_{\mathbf{s} \in \mathcal{S}_n} \widetilde{p}_{\text{POISSWOR}}(\mathbf{s}, \boldsymbol{\lambda} + c\mathbf{1})}.$$

From Expression (5.8), page 76, we have

$$\widetilde{\pi}_k = \frac{\exp(\lambda_k + c)}{1 + \exp(\lambda_k + c)}.$$

The constant c can be chosen freely, but it can be convenient for c that

$$\sum_{k \in U} \widetilde{\pi}_k = n.$$

Thus,

$$\lambda_k + c = \log \frac{\widetilde{\pi}_k}{1 - \widetilde{\pi}_k}.$$

Define now $\boldsymbol{\pi}$ as a function of $\widetilde{\boldsymbol{\pi}}$, that is denoted $\boldsymbol{\psi}(\widetilde{\boldsymbol{\pi}}, n)$:

$$\boldsymbol{\psi}(\widetilde{\boldsymbol{\pi}}, n) = \boldsymbol{\pi}(\boldsymbol{\lambda}, \mathcal{S}_n) = \frac{\sum_{\mathbf{s} \in \mathcal{S}_n} \mathbf{s} \exp \boldsymbol{\lambda}'\mathbf{s}}{\sum_{\mathbf{s} \in \mathcal{S}_n} \exp \boldsymbol{\lambda}'\mathbf{s}} = \frac{\sum_{\mathbf{s} \in \mathcal{S}_n} \mathbf{s} \prod_{k \in U} \left(\frac{\widetilde{\pi}_k}{1 - \widetilde{\pi}_k}\right)^{s_k}}{\sum_{\mathbf{s} \in \mathcal{S}_n} \prod_{k \in U} \left(\frac{\widetilde{\pi}_k}{1 - \widetilde{\pi}_k}\right)^{s_k}},$$

and $\boldsymbol{\pi}(\boldsymbol{\lambda}, \mathcal{S}_0) = \mathbf{0}$. Expression (5.13) becomes:

$$\psi_k(\widetilde{\boldsymbol{\pi}}, n) = n \frac{\frac{\widetilde{\pi}_k}{1 - \widetilde{\pi}_k}\{1 - \psi_k(\widetilde{\boldsymbol{\pi}}, n-1)\}}{\sum_{\ell \in U} \frac{\widetilde{\pi}_\ell}{1 - \widetilde{\pi}_\ell}\{1 - \psi_\ell(\widetilde{\boldsymbol{\pi}}, n-1)\}}. \tag{5.15}$$

If the inclusion probabilities $\boldsymbol{\pi}$ (such that $\sum_{k \in U} \pi_k = n$) are given, Deville (2000) and Chen (2000) have proposed to solve the equation

$$\boldsymbol{\psi}(\widetilde{\boldsymbol{\pi}}, n) = \boldsymbol{\pi},$$

in $\widetilde{\boldsymbol{\pi}}$ by the Newton method, which gives the algorithm

$$\widetilde{\boldsymbol{\pi}}^{(i)} = \widetilde{\boldsymbol{\pi}}^{(i-1)} + \left| \frac{\partial \boldsymbol{\psi}(\widetilde{\boldsymbol{\pi}}, n)}{\partial \widetilde{\boldsymbol{\pi}}'} \right|_{\widetilde{\boldsymbol{\pi}} = \widetilde{\boldsymbol{\pi}}^{(i-1)}}^{-1} \left[\boldsymbol{\pi} - \boldsymbol{\psi}(\widetilde{\boldsymbol{\pi}}^{(i-1)}, n) \right],$$

where $i = 1, 2, \ldots$ and with $\widetilde{\boldsymbol{\pi}}^{(0)} = \boldsymbol{\pi}$. Let $\boldsymbol{\Delta}$ be the matrix of $\pi_{k\ell} - \pi_k \pi_\ell$'s, Deville (2000) has pointed out that the matrix

$$\frac{\partial \boldsymbol{\psi}(\widetilde{\boldsymbol{\pi}}, n)}{\partial \widetilde{\boldsymbol{\pi}}'} = \boldsymbol{\Delta} \left\{ \mathrm{diag}[\widetilde{\pi}_1(1 - \widetilde{\pi}_1) \cdots \widetilde{\pi}_k(1 - \widetilde{\pi}_k) \cdots \widetilde{\pi}_N(1 - \widetilde{\pi}_N)] \right\}^{-1}$$

is very close to the identity matrix, which allows simplifying the algorithm significantly. Finally, we can use

$$\widetilde{\boldsymbol{\pi}}^{(i)} = \widetilde{\boldsymbol{\pi}}^{(i-1)} + \boldsymbol{\pi} - \boldsymbol{\psi}(\widetilde{\boldsymbol{\pi}}^{(i-1)}, n), \tag{5.16}$$

which allows us quickly to derive $\widetilde{\boldsymbol{\pi}}$ and thus $\lambda_k = \log(\widetilde{\pi}_k/(1 - \widetilde{\pi}_k))$ from $\boldsymbol{\pi}$. Aires (1999) has used the same algorithm with another definition of $\boldsymbol{\psi}$ that gives a slower method. The number of operations needed to compute $\widetilde{\pi}_k$ is about $O(N \times n \times$ number of iterations). Table 5.3 presents an example of the computation of $\boldsymbol{\lambda}$ from a given $\boldsymbol{\pi}$ by means of the Newton method.

Table 5.3. Computation of $\boldsymbol{\lambda}$ from a given $\boldsymbol{\pi}$ by means of the Newton method

k	π_k	$\widetilde{\pi}_k$	$\lambda_k = \log \frac{\widetilde{\pi}_k}{1 - \widetilde{\pi}_k}$	rescaled λ_k
1	0.07	0.1021	-2.1743	-2.1514
2	0.17	0.2238	-1.2436	-1.2206
3	0.41	0.4417	-0.2342	-0.2112
4	0.61	0.5796	0.3212	0.3442
5	0.83	0.7794	1.2619	1.2848
6	0.91	0.8734	1.9313	1.9543
	3	3	-0.1377	0

A variant of recursive relation (5.16) consists of working on

$$\mathrm{logit}\, \pi_k = \log \frac{\pi_k}{1 - \pi_k}.$$

The general step of the Newton method becomes

$$\mathrm{logit}\, \widetilde{\boldsymbol{\pi}}^{(i)} = \mathrm{logit}\, \widetilde{\boldsymbol{\pi}}^{(i-1)} + \left| \frac{\partial \mathrm{logit}\, \boldsymbol{\psi}(\widetilde{\boldsymbol{\pi}}, n)}{\partial \mathrm{logit}\, \widetilde{\boldsymbol{\pi}}} \right|_{\widetilde{\boldsymbol{\pi}} = \widetilde{\boldsymbol{\pi}}^{(i-1)}} [\mathrm{logit}\, \boldsymbol{\pi} - \mathrm{logit}\, \boldsymbol{\psi}(\widetilde{\boldsymbol{\pi}}^{(i-1)}, n)]. \tag{5.17}$$

Note that logit $\widetilde{\boldsymbol{\pi}} = \boldsymbol{\lambda}$. Moreover

$$\frac{\partial\text{logit }\boldsymbol{\psi}(\widetilde{\boldsymbol{\pi}}, n)}{\partial\text{logit }\widetilde{\boldsymbol{\pi}}} = \frac{\partial\text{logit }\boldsymbol{\pi}(\boldsymbol{\lambda}, \mathcal{S}_n)}{\partial\boldsymbol{\lambda}}$$

$$= \boldsymbol{\Delta} \text{ diag}\left(\widetilde{\pi}_1(1 - \widetilde{\pi}_1) \ \ldots \ \widetilde{\pi}_k(1 - \widetilde{\pi}_k) \ \ldots \ \widetilde{\pi}_N(1 - \widetilde{\pi}_N)\right)^{-1}$$

Again, $\boldsymbol{\Delta} \text{ diag}\left(\widetilde{\pi}_1(1 - \widetilde{\pi}_1) \ \ldots \ \widetilde{\pi}_k(1 - \widetilde{\pi}_k) \ \ldots \ \widetilde{\pi}_N(1 - \widetilde{\pi}_N)\right)^{-1}$ is close to the identity matrix. Expression (5.17) can be approximated by

$$\boldsymbol{\lambda}^{(i)} = \boldsymbol{\lambda}^{(i-1)} + \text{logit }\boldsymbol{\pi} - \text{logit }\boldsymbol{\pi}(\boldsymbol{\lambda}^{(i-1)}, \mathcal{S}_n), \tag{5.18}$$

with the initial value $\boldsymbol{\lambda}^{(0)} = \text{logit }\boldsymbol{\pi}$.

5.6.4 Joint Inclusion Probabilities

The joint inclusion probabilities can also be computed. The following result has been presented in Chen et al. (1994), Aires (1999), Chen and Liu (1997), and Chen (2000).

Result 23. *If*

$$\pi_{k\ell}(\boldsymbol{\lambda}, \mathcal{S}_n) = \frac{\sum_{\mathbf{s}\in\mathcal{S}_n} s_k s_\ell \exp\boldsymbol{\lambda}'\mathbf{s}}{\sum_{\mathbf{s}\in\mathcal{S}_n} \exp\boldsymbol{\lambda}'\mathbf{s}},$$

and $\boldsymbol{\pi}(\boldsymbol{\lambda}, \mathcal{S}_n)$ *is given in (5.11) then for* $k \neq \ell$:

$$(i) \ \ \pi_{k\ell}(\boldsymbol{\lambda}, \mathcal{S}_{n+1}) = [\pi_k(\boldsymbol{\lambda}, \mathcal{S}_n) - \pi_{k\ell}(\boldsymbol{\lambda}, \mathcal{S}_n)] \, (\exp\lambda_\ell)\frac{\exp\alpha(\boldsymbol{\lambda}, \mathcal{S}_n)}{\exp\alpha(\boldsymbol{\lambda}, \mathcal{S}_{n+1})}, \tag{5.19}$$

$$(ii) \ \ \pi_{k\ell}(\boldsymbol{\lambda}, \mathcal{S}_n) = \frac{\pi_k(\boldsymbol{\lambda}, \mathcal{S}_n)\exp\lambda_\ell - \pi_\ell(\boldsymbol{\lambda}, \mathcal{S}_n)\exp\lambda_k}{\exp\lambda_\ell - \exp\lambda_k}, \ \ \textit{if } \lambda_k \neq \lambda_\ell, \tag{5.20}$$

$$(iii) \ \ \pi_{k\ell}(\boldsymbol{\lambda}, \mathcal{S}_{n+1}) = \frac{\pi_\ell(\boldsymbol{\lambda}, \mathcal{S}_{n+1})[\pi_k(\boldsymbol{\lambda}, \mathcal{S}_n) - \pi_{k\ell}(\boldsymbol{\lambda}, \mathcal{S}_n)]}{1 - \pi_\ell(\boldsymbol{\lambda}, \mathcal{S}_n)}, \tag{5.21}$$

$$(iv) \ \ \pi_{k\ell}(\boldsymbol{\lambda}, \mathcal{S}_{n+1}) = -\sum_{i=1}^{n}\pi_k(\boldsymbol{\lambda}, \mathcal{S}_i)\prod_{j=i}^{n}\frac{-\pi_\ell(\boldsymbol{\lambda}, \mathcal{S}_{j+1})}{1 - \pi_\ell(\boldsymbol{\lambda}, \mathcal{S}_j)}. \tag{5.22}$$

Proof. (i)

$$[\pi_k(\boldsymbol{\lambda}, \mathcal{S}_n) - \pi_{k\ell}(\boldsymbol{\lambda}, \mathcal{S}_n)]\exp\lambda_\ell$$

$$= \left[\sum_{\substack{\mathbf{s}\in\mathcal{S}_n \\ s_k=1}} p_{\text{CPS}}(\mathbf{s}, \boldsymbol{\lambda}, n) - \sum_{\substack{\mathbf{s}\in\mathcal{S}_n \\ s_k=1 \\ s_\ell=1}} p_{\text{CPS}}(\mathbf{s}, \boldsymbol{\lambda}, n)\right]\exp\lambda_\ell$$

$$= \sum_{\substack{\mathbf{s}\in\mathcal{S}_n \\ s_k=1 \\ s_\ell=0}} p_{\text{CPS}}(\mathbf{s}, \boldsymbol{\lambda}, n)\exp\lambda_\ell = \frac{\sum_{\substack{\mathbf{s}\in\mathcal{S}_n \\ s_k=1 \\ s_\ell=0}} \exp(\boldsymbol{\lambda}'\mathbf{s} + \lambda_\ell)}{\exp\alpha(\boldsymbol{\lambda}, \mathcal{S}_n)}$$

$$= \frac{\sum_{\substack{s \in \mathcal{S}_{n+1} \\ s_k=1 \\ s_\ell=1}} \exp(\boldsymbol{\lambda}'\mathbf{s})}{\exp \alpha \left(\boldsymbol{\lambda}, \mathcal{S}_{n+1}\right)} \frac{\exp \alpha \left(\boldsymbol{\lambda}, \mathcal{S}_{n+1}\right)}{\exp \alpha \left(\boldsymbol{\lambda}, \mathcal{S}_n\right)} = \pi_{k\ell}(\boldsymbol{\lambda}, \mathcal{S}_{n+1}) \frac{\exp \alpha \left(\boldsymbol{\lambda}, \mathcal{S}_{n+1}\right)}{\exp \alpha \left(\boldsymbol{\lambda}, \mathcal{S}_n\right)}.$$

(ii) By symmetry, by Expression (5.19), we can also write

$$\pi_{k\ell}(\boldsymbol{\lambda}, \mathcal{S}_{n+1}) = [\pi_\ell(\boldsymbol{\lambda}, \mathcal{S}_n) - \pi_{k\ell}(\boldsymbol{\lambda}, \mathcal{S}_n)] \left(\exp \lambda_k\right) \frac{\exp \alpha \left(\boldsymbol{\lambda}, \mathcal{S}_n\right)}{\exp \alpha \left(\boldsymbol{\lambda}, \mathcal{S}_{n+1}\right)}. \quad (5.23)$$

By subtracting (5.19) from (5.23), we get

$$[\pi_k(\boldsymbol{\lambda}, \mathcal{S}_n) - \pi_{k\ell}(\boldsymbol{\lambda}, \mathcal{S}_n)] \exp \lambda_k - [\pi_\ell(\boldsymbol{\lambda}, \mathcal{S}_n) - \pi_{k\ell}(\boldsymbol{\lambda}, \mathcal{S}_n)] \exp \lambda_\ell = 0,$$

which gives (5.20).

(iii) By summing (5.19) over $k \in U \backslash \{\ell\}$, we get

$$\pi_\ell(\boldsymbol{\lambda}, \mathcal{S}_{n+1}) = [1 - \pi_\ell(\boldsymbol{\lambda}, \mathcal{S}_n)] \left(\exp \lambda_\ell\right) \frac{\exp \alpha \left(\boldsymbol{\lambda}, \mathcal{S}_n\right)}{\exp \alpha \left(\boldsymbol{\lambda}, \mathcal{S}_{n+1}\right)}. \quad (5.24)$$

With Expressions (5.19) and (5.24), we get (5.21).

(iv) Expression (5.22) satisfies the recurrence equation (5.21).   □

If all the λ_k's are distinct, the joint inclusion probabilities can be computed very quickly by Expression (5.20). If some inclusion probabilities are equal, Aires (1999) pointed out that, since

$$\sum_{\substack{k \in U \\ k \neq \ell}} \pi_{k\ell} = \pi_\ell(n-1),$$

we have for all $k \neq \ell$ such that $\lambda_k = \lambda_\ell$

$$\pi_{k\ell} = \frac{1}{\operatorname{card} \left\{ k \in U \backslash \{k\} | \lambda_k = \lambda_\ell \right\}} \left[\pi_\ell(n-1) - \sum_{\substack{k \in U \\ \ell \neq \ell, \lambda_k \neq \lambda_\ell}} \pi_{k\ell} \right].$$

5.6.5 Joint Inclusion Probabilities: Deville's Technique

Once $\boldsymbol{\lambda}$ is derived from $\boldsymbol{\pi}$, Deville (2000) has also given a fast method to compute the joint inclusion probabilities.

Result 24. *The joint inclusion probabilities of a CPS satisfy the recursive equation: (size equal to n)*

$$\pi_{k\ell}(\boldsymbol{\lambda}, \mathcal{S}_n)$$
$$= \frac{n(n-1) \exp \lambda_k \exp \lambda_\ell \left[1 - \pi_k(\boldsymbol{\lambda}, \mathcal{S}_{n-2}) - \pi_\ell(\boldsymbol{\lambda}, \mathcal{S}_{n-2}) + \pi_{k\ell}(\boldsymbol{\lambda}, \mathcal{S}_{n-2})\right]}{\sum_{i \in U} \sum_{\substack{j \in U \\ i \neq j}} \exp \lambda_i \exp \lambda_j \left[1 - \pi_i(\boldsymbol{\lambda}, \mathcal{S}_{n-2}) - \pi_j(\boldsymbol{\lambda}, \mathcal{S}_{n-2}) + \pi_{ij}(\boldsymbol{\lambda}, \mathcal{S}_{n-2})\right]},$$

(5.25)

with $\pi_{k\ell}(\boldsymbol{\lambda}, 0) = 0$, $\pi_{k\ell}(\boldsymbol{\lambda}, 1) = 0$, for all $k, \ell \in U, k \neq \ell$.

Proof. We have

$$\pi_{k\ell}(\boldsymbol{\lambda}, \mathcal{S}_n)$$

$$= \sum_{\mathbf{s}\in\mathcal{S}_n} s_k s_\ell p_{\mathrm{EXP}}(\mathbf{s}, \boldsymbol{\lambda}, \mathcal{S}_n) = \frac{\sum_{\mathbf{s}\in\mathcal{S}_n} s_k s_\ell \exp \boldsymbol{\lambda}'\mathbf{s}}{\alpha(\boldsymbol{\lambda}, \mathcal{S}_n)} = \sum_{\substack{\mathbf{s}\in\mathcal{S}_n \\ s_k=s_\ell=1}} \frac{\exp \boldsymbol{\lambda}'\mathbf{s}}{\alpha(\boldsymbol{\lambda}, \mathcal{S}_n)}$$

$$= \frac{\exp \lambda_k \exp \lambda_\ell \sum_{\substack{\mathbf{s}\in\mathcal{S}_{n-2} \\ s_k=s_\ell=0}} \exp \boldsymbol{\lambda}'\mathbf{s}}{\alpha(\boldsymbol{\lambda}, \mathcal{S}_n)}$$

$$= \frac{\exp \lambda_k \exp \lambda_\ell}{\alpha(\boldsymbol{\lambda}, \mathcal{S}_n)} \left(\sum_{\mathbf{s}\in\mathcal{S}_{n-2}} \exp \boldsymbol{\lambda}'\mathbf{s} - \sum_{\substack{\mathbf{s}\in\mathcal{S}_{n-2} \\ s_k=1}} \exp \boldsymbol{\lambda}'\mathbf{s} \right.$$

$$\left. - \sum_{\substack{\mathbf{s}\in\mathcal{S}_{n-2} \\ s_\ell=1}} \exp \boldsymbol{\lambda}'\mathbf{s} + \sum_{\substack{\mathbf{s}\in\mathcal{S}_{n-2} \\ s_k=1, s_\ell=1}} \exp \boldsymbol{\lambda}'\mathbf{s} \right)$$

$$= \exp \lambda_k \exp \lambda_\ell \left[1 - \pi_k(\boldsymbol{\lambda}, \mathcal{S}_{n-2}) - \pi_\ell(\boldsymbol{\lambda}, \mathcal{S}_{n-2}) + \pi_{k\ell}(\boldsymbol{\lambda}, \mathcal{S}_{n-2}) \right] \frac{\alpha(\boldsymbol{\lambda}, \mathcal{S}_{n-2})}{\alpha(\boldsymbol{\lambda}, \mathcal{S}_n)}.$$

Because

$$\sum_{k\in U} \sum_{\substack{\ell\in U \\ k\neq\ell}} \pi_{k\ell}(\boldsymbol{\lambda}, \mathcal{S}_n) = n(n-1),$$

we finally get Expression (5.25). □

Example 10. With the CPS method, for $N = 6, n = 3$, and

$$\boldsymbol{\pi} = (0.07\ 0.17\ 0.41\ 0.61\ 0.83\ 0.91)',$$

vector $\boldsymbol{\lambda}$ is computed in Table 5.3, page 83. The matrix of joint inclusion probabilities is given by

$$\boldsymbol{\Pi} = \begin{pmatrix} 0.07 & 0.0049 & 0.0130 & 0.0215 & 0.0447 & 0.0559 \\ 0.0049 & 0.17 & 0.0324 & 0.0537 & 0.1113 & 0.1377 \\ 0.0130 & 0.0324 & 0.41 & 0.1407 & 0.2888 & 0.3452 \\ 0.0215 & 0.0537 & 0.1407 & 0.61 & 0.4691 & 0.5351 \\ 0.0447 & 0.1113 & 0.2888 & 0.4691 & 0.83 & 0.7461 \\ 0.0559 & 0.1377 & 0.3452 & 0.5351 & 0.7461 & 0.91 \end{pmatrix}. \tag{5.26}$$

Qualité (2005) has shown that conditional Poisson sampling provides always a smaller variance than multinomial sampling with the same sample size and the same expectation of the random sample.

5.6.6 Computation of $\alpha(\boldsymbol{\lambda}, \mathcal{S}_n)$

The computation of the normalizing constant can be done recursively and is directly linked to Expressions (5.9) and (5.10), page 78 (see also Chen and Liu, 1997; Aires, 1999; Sampford, 1967, p. 507). Define

$$U_i^j = \{i, i+1, \ldots, j-1, j\},$$

$$\mathcal{S}_z(U_i^j) = \left\{\mathbf{s} \in \mathcal{S}_z | s_k = 0 \text{ if } k \notin U_i^j\right\},$$

with $z \leq j - i + 1$,

$$\alpha\left(\boldsymbol{\lambda}, \mathcal{S}_z(U_i^j)\right) = \log \sum_{\mathbf{s} \in \mathcal{S}_z(U_i^j)} \exp \boldsymbol{\lambda}'\mathbf{s}.$$

These particular cases are obvious:

$$\alpha\left[\boldsymbol{\lambda}, \mathcal{S}_1(U_i^j)\right] = \log \sum_{k=i}^{j} \exp \lambda_k, \tag{5.27}$$

$$\alpha\left[\boldsymbol{\lambda}, \mathcal{S}_{\text{card}U_i^j}(U_i^j)\right] = \sum_{k=i}^{j} \lambda_k. \tag{5.28}$$

The general case can be computed recursively by means of the following results:

(i) $\exp \alpha\left[\boldsymbol{\lambda}, \mathcal{S}_z(U_i^N)\right]$

$$= (\exp \lambda_i) \exp \alpha\left[\boldsymbol{\lambda}, \mathcal{S}_{z-1}(U_{i+1}^N)\right] + \exp \alpha\left[\boldsymbol{\lambda}, \mathcal{S}_z(U_{i+1}^N)\right] \tag{5.29}$$

(ii) $\exp \alpha\left[\boldsymbol{\lambda}, \mathcal{S}_z(U_1^j)\right]$

$$= (\exp \lambda_j) \exp \alpha\left[\boldsymbol{\lambda}, \mathcal{S}_{z-1}(U_1^{j-1})\right] + \exp \alpha\left[\boldsymbol{\lambda}, \mathcal{S}_z(U_1^{j-1})\right].$$

An example of the recursive computation of $\alpha[\boldsymbol{\lambda}, \mathcal{S}_z(U_k^N)]$ by means of Expressions (5.27), (5.28), and (5.29) is given in Table 5.4.

Table 5.4. Recursive computation of $\exp \alpha[\boldsymbol{\lambda}, \mathcal{S}_z(U_k^N)]$: the computation is made from the bottom to the top of the table by means of Expressions (5.27), (5.28), and (5.29)

k	λ_k	$\exp(\lambda_k)$	$z = 1$	2	3	4	5	6
1	-2.151	0.116	13.305	55.689	89.990	59.182	14.349	1.000
2	-1.221	0.295	13.188	54.155	83.691	49.448	8.597	
3	-0.211	0.810	12.893	50.351	68.835	29.138		
4	0.344	1.411	12.084	40.568	35.991			
5	1.285	3.614	10.673	25.511				
6	1.954	7.059	7.059					

Chen et al. (1994), Chen and Liu (1997), and Chen (1998, 2000) have given several other relations on $\alpha\left(\boldsymbol{\lambda}, \mathcal{S}_n\right)$.

Result 25. *For any $V \subset U$ and $1 \leq n \leq \operatorname{card} V$, then*

(i) $\exp \alpha \left[\boldsymbol{\lambda}, \mathcal{S}_n(V)\right] = \dfrac{1}{n} \displaystyle\sum_{k \in V} \exp \left\{\lambda_k + \alpha \left[\boldsymbol{\lambda}, \mathcal{S}_{n-1}(V \backslash \{k\})\right]\right\},$

(ii) $\exp \alpha \left[\boldsymbol{\lambda}, \mathcal{S}_n(V)\right] = \dfrac{1}{\operatorname{card} V - n} \displaystyle\sum_{k \in V} \exp \left\{\alpha \left[\boldsymbol{\lambda}, \mathcal{S}_n(V \backslash \{k\})\right]\right\},$

(iii) $\exp \alpha \left[\boldsymbol{\lambda}, \mathcal{S}_n(V)\right] = \displaystyle\sum_{\ell=1}^{n} \exp \left\{\alpha \left[\boldsymbol{\lambda}, \mathcal{S}_\ell(V)\right] + \alpha \left[\boldsymbol{\lambda}, \mathcal{S}_{n-\ell}(U \backslash V)\right]\right\},$

(iv) $\exp \alpha \left[\boldsymbol{\lambda}, \mathcal{S}_n(V)\right] = \dfrac{1}{n} \displaystyle\sum_{\ell=1}^{n} (-1)^{\ell+1} \left[\sum_{j \in V} \exp(\ell \lambda_j)\right] \exp \alpha \left[\boldsymbol{\lambda}, \mathcal{S}_{n-\ell}(V)\right].$

Proof.

(i) $\quad \dfrac{1}{n} \displaystyle\sum_{k \in V} \exp \left\{\lambda_k + \alpha \left[\boldsymbol{\lambda}, \mathcal{S}_{n-1}(V \backslash \{k\})\right]\right\}$

$= \dfrac{1}{n} \displaystyle\sum_{k \in V} (\exp \lambda_k) \sum_{\mathbf{s} \in \mathcal{S}_{n-1}(V \backslash \{k\})} \exp \boldsymbol{\lambda}' \mathbf{s}$

$= \dfrac{1}{n} \displaystyle\sum_{k \in V} \sum_{\mathbf{s} \in \mathcal{S}_n(V)|s_k=1} \exp \boldsymbol{\lambda}' \mathbf{s} = \exp \alpha \left[\boldsymbol{\lambda}, \mathcal{S}_n(V)\right].$

(ii) $\quad \dfrac{1}{\operatorname{card} V - n} \displaystyle\sum_{k \in V} \exp \left\{\alpha \left[\boldsymbol{\lambda}, \mathcal{S}_n(V \backslash \{k\})\right]\right\}$

$= \dfrac{1}{\operatorname{card} V - n} \displaystyle\sum_{k \in V} \sum_{\mathbf{s} \in \mathcal{S}_n(V \backslash \{k\})} \exp \boldsymbol{\lambda}' \mathbf{s}$

$= \displaystyle\sum_{\mathbf{s} \in \mathcal{S}_n(V)} \exp \boldsymbol{\lambda}' \mathbf{s} = \exp \alpha \left[\boldsymbol{\lambda}, \mathcal{S}_n(V)\right].$

(iii) $\quad \displaystyle\sum_{\ell=1}^{\max(n, \operatorname{card} V)} \exp \left\{\alpha \left(\boldsymbol{\lambda}, \mathcal{S}_\ell(V)\right) + \alpha \left[\boldsymbol{\lambda}, \mathcal{S}_{n-\ell}(U \backslash V)\right]\right\}$

$= \displaystyle\sum_{\ell=1}^{\max(n, \operatorname{card} V)} \sum_{\mathbf{s} \in \mathcal{S}_\ell(V)} \exp \boldsymbol{\lambda}' \mathbf{s} \sum_{\mathbf{r} \in \mathcal{S}_{n-\ell}(U \backslash V)} \exp \boldsymbol{\lambda}' \mathbf{r}$

$= \displaystyle\sum_{\mathbf{s} \in \mathcal{S}_n(U)} \exp \boldsymbol{\lambda}' \mathbf{s} = \exp \alpha \left[\boldsymbol{\lambda}, \mathcal{S}_n(U)\right]$

(iv) Chen et al. (1994) have given an independent proof of this relation. Result 25 *(iv)* can, however, be derived by summing over all $k \in U$, Expression (5.14) of Result 22, page 81. $\qquad \square$

5.6.7 Poisson Rejective Procedure for CPS

The CPS design can be obtained by conditioning a Poisson sampling design on a fixed sample size.

Result 26.

$$p_{\mathrm{EXP}}(\mathbf{s}, \boldsymbol{\lambda}, \mathcal{S}_n) = p_{\mathrm{EXP}}(\mathbf{s}, \boldsymbol{\lambda} + c\mathbf{1}, \mathcal{S}|n(\mathbf{S}) = n) = \frac{p_{\mathrm{EXP}}(\mathbf{s}, \boldsymbol{\lambda} + c\mathbf{1}, \mathcal{S})}{\sum_{\mathbf{s} \in \mathcal{S}_n} p_{\mathrm{EXP}}(\mathbf{s}, \boldsymbol{\lambda} + c\mathbf{1}, \mathcal{S})},$$

for all $\mathbf{s} \in \mathcal{S}_n$ and $c \in \mathbb{R}$.

The proof is obvious. Note that a CPS with parameter $\boldsymbol{\lambda}$ can be obtained by conditioning an infinity of Poisson sampling design with parameter $\boldsymbol{\lambda} + c\mathbf{1}$, for any $c \in \mathbb{R}$. Result 26 allows defining the rejective method described in Algorithm 5.5.

Algorithm 5.5 Poisson rejective procedure for CPS

---

1. Compute $\boldsymbol{\lambda}$ and $\widetilde{\boldsymbol{\pi}}$ from $\boldsymbol{\pi}$ by means of the Newton method as written in (5.16).
2. Select a random sample $\widetilde{\mathbf{S}}$ with the Poisson design $p_{\mathrm{EXP}}(\widetilde{\mathbf{s}}, \boldsymbol{\lambda} + c\mathbf{1}, \mathcal{S})$.
3. IF the sample size is not equal to n GOTO STEP 2, ELSE STOP; ENDIF.

Constant c can be any real number, but c should be chosen to minimize $1/\Pr[n(\widetilde{\mathbf{S}}) = n]$. A convenient solution consists of using a constant c such that

$$\sum_{k \in U} \widetilde{\pi}_k = \sum_{k \in U} \frac{\exp(\lambda_k + c)}{1 + \exp(\lambda_k + c)} = n. \tag{5.30}$$

Note that recursive relation (5.16) provides $\widetilde{\pi}_k$'s that directly have such properties. If the λ_k's are such that $\sum_{k \in U} \lambda_k = 0$, then an interesting solution could be $c = \log(n/N - n)$, which provides a valid approximation for (5.30).

The expected number of iterations is $1/\Pr\left[n(\widetilde{\mathbf{S}}) = n\right]$. Contrary to popular belief, this quantity is not very large and does not increase very much with the population size. If $\widetilde{n} = n(\widetilde{\mathbf{S}})$ denotes the random sample size of the Poisson design, then

$$\mathrm{E}(\widetilde{n}) = \sum_{k \in U} \widetilde{\pi}_k = n, \quad \text{and} \quad \mathrm{var}(\widetilde{n}) = \sum_{k \in U} \widetilde{\pi}_k(1 - \widetilde{\pi}_k) \leq N \frac{n}{N} \frac{N - n}{N}.$$

Asymptotically, $\widetilde{n}$ has a normal distribution and

$$\widetilde{n} \sim N\left(n, n\frac{N - n}{N}\right),$$

and thus

$$\Pr(\widetilde{n} = n) \approx \Pr\left[-\frac{1}{2} \leq N\left(n, n\frac{N-n}{N}\right) \leq \frac{1}{2}\right]$$

$$\approx \Pr\left[-\frac{1}{2\sqrt{n\frac{N-n}{N}}} \leq N(0,1) \leq \frac{1}{2\sqrt{n\frac{N-n}{N}}}\right]$$

$$\approx f(0)\frac{1}{\sqrt{n\frac{N-n}{N}}} = \frac{1}{\sqrt{2\pi n\frac{N-n}{N}}},$$

where $N(0,1)$ is a centered reduced normal variable and $f(.)$ is its density. The expected number of operations needed to get the sample is about

$$\frac{N}{\Pr\left[n(\widetilde{\mathbf{S}}) = n\right]} \approx N\sqrt{2\pi n\frac{N-n}{N}},$$

which is very fast for a sampling method with unequal probabilities.

5.6.8 Rejective Procedure with Multinomial Design for CPS

We have the following result.

Result 27.

$$p_{\mathrm{EXP}}(\mathbf{s}, \boldsymbol{\lambda}, \mathcal{S}_n) = p_{\mathrm{EXP}}(\mathbf{s}, \boldsymbol{\lambda}, \mathcal{R}_n | r(\mathbf{S}) = \mathbf{s}) = \frac{p_{\mathrm{EXP}}(\mathbf{s}, \boldsymbol{\lambda}, \mathcal{R}_n)}{\sum_{\mathbf{s} \in \mathcal{S}_n} p_{\mathrm{EXP}}(\mathbf{s}, \boldsymbol{\lambda}, \mathcal{R}_n)},$$

for all $\mathbf{s} \in \mathcal{S}_n$, *where* $r(.)$ *is the reduction function defined in Section 2.7, page 14.*

The proof is obvious. The rejective method is defined in Algorithm 5.6.

Algorithm 5.6 Rejective procedure with multinomial design for CPS

---

1. Compute $\boldsymbol{\lambda}$ from $\boldsymbol{\pi}$ by means of the Newton method as written in Expression (5.16).
2. Select a random sample $\mathbf{S}$ with a multinomial design $p_{\mathrm{EXP}}(\mathbf{s}, \boldsymbol{\lambda}, \mathcal{R}_n)$.
3. IF one of the $\widetilde{s}_k > 1, k \in U$ THEN GOTO STEP 2; ELSE STOP; ENDIF.

This algorithm is presented in Brewer and Hanif (1983, Carrol-Hartley rejective procedure 14, page 31, procedure 28-31, page 40) and was also discussed in Rao (1963a), Carroll and Hartley (1964), Hájek (1964), Rao and Bayless (1969), Bayless and Rao (1970), and Cassel et al. (1993), but the way to rapidly calculate the exact working probabilities μ_k/n was not yet known.

A very simple implementation given in Algorithm 5.7 can be done from Algorithm 5.2. The rejective procedure with replacement can be faster than the Poisson rejective procedure when n/N is small.

Algorithm 5.7 Rejective procedure with sequential sampling with replacement for CPS

DEFINITION k: INTEGER;

1. Compute $\boldsymbol{\lambda}$ from $\boldsymbol{\pi}$ by means of the Newton method as written in (5.16);
2. Compute $\mu_k = \dfrac{n \exp \lambda_k}{\sum_{\ell \in U} \exp \lambda_\ell}$;
3. FOR $k = 1, \dots, N$ DO

   Select the kth unit s_k times according to the binomial distribution
   $$\mathcal{B}\left(n - \sum_{\ell=1}^{k-1} s_\ell, \frac{\mu_k}{n - \sum_{\ell=1}^{k-1} \mu_\ell}\right);$$

   IF $s_k \geq 2$ THEN GOTO STEP 3; ENDIF;
   ENDFOR.

5.6.9 Sequential Procedure for CPS

The sequential procedure for CPS was proposed by Chen and Liu (1997, procedure 2) and Deville (2000) and is based on the following result.

Result 28.

$$
\begin{aligned}
q_{k(n-j)} &= \mathrm{E}(S_k | S_{k-1} = s_{k-1}, \dots, S_1 = s_1) \\
&= (\exp \lambda_k) \frac{\exp \alpha \left[\boldsymbol{\lambda}, \mathcal{S}_{n-j-1}(U_{k+1}^N)\right]}{\exp \alpha \left[\boldsymbol{\lambda}, \mathcal{S}_{n-j}(U_k^N)\right]},
\end{aligned}
\tag{5.31}
$$

where

$$j = \sum_{i=1}^{k-1} s_i,$$

and the notation is the same as in Section 5.6.6.

Proof.

$$\mathrm{E}(S_k | S_{k-1} = s_{k-1}, \dots, S_1 = s_1) = \frac{\displaystyle\sum_{\mathbf{s} \in \mathcal{S}_{n-j}(U_k^N)} s_k \exp \boldsymbol{\lambda}'\mathbf{s}}{\displaystyle\sum_{\mathbf{s} \in \mathcal{S}_{n-j}(U_k^N)} \exp \boldsymbol{\lambda}'\mathbf{s}}$$

$$= (\exp \lambda_k) \frac{\displaystyle\sum_{\mathbf{s} \in \mathcal{S}_{n-j-1}(U_{k+1}^N)} \exp \boldsymbol{\lambda}'\mathbf{s}}{\displaystyle\sum_{\mathbf{s} \in \mathcal{S}_{n-j}(U_k^N)} \exp \boldsymbol{\lambda}'\mathbf{s}} = (\exp \lambda_k) \frac{\exp \alpha \left[\boldsymbol{\lambda}, \mathcal{S}_{n-j-1}(U_{k+1}^N)\right]}{\exp \alpha \left[\boldsymbol{\lambda}, \mathcal{S}_{n-j}(U_k^N)\right]}. \quad \square$$

Note that the probability of selecting unit k given that $S_{k-1} = s_{k-1}, \dots, S_1 = s_1$ depends only on the number of selected units; that is,

$$\mathrm{Pr}(S_k = s_k | S_{k-1} = s_{k-1}, \dots, S_1 = s_1) = \mathrm{Pr}\left(S_k = s_k \left| \sum_{i=1}^{k-1} S_i = j\right.\right).$$

Example 11. For $N = 6, n = 3$, and $\pi = (0.07\ 0.17\ 0.41\ 0.61\ 0.83\ 0.91)'$, vector λ is computed in Table 5.3, page 83. Table 5.4, page 87, gives the values $\exp \alpha[\lambda, \mathcal{S}_z(U_k^N)]$ that are computed recursively from λ. Finally, in Table 5.5, the $q_{k(n-j)}$ are computed directly from the table of $\exp \alpha[\lambda, \mathcal{S}_z(U_k^N)]$, which allows a quick implementation of the CPS design.

Table 5.5. Computation of q_{kj} from Table 5.4, page 87

$j = 1$	2	3	4	5	6
$k = 1$ 0.009	0.028	0.070	0.164	0.401	1
2 0.022	0.070	0.178	0.411	1	
3 0.063	0.194	0.477	1		
4 0.117	0.371	1			
5 0.339	1				
6 1					

In order to select a sample with the sampling design $p_{\mathrm{EXP}}(\mathbf{s}, \mathcal{S}_n, \lambda)$, the standard sequential Algorithm 3.2, page 34, gives Algorithm 5.8 (Deville, 2000; Chen and Liu, 1997, procedure 3).

Algorithm 5.8 Sequential procedure for CPS

1. Compute λ from π by means of the Newton method given in Expression (5.16).
2. FOR $k = 0, \ldots, N$, DO FOR $z = 1, \ldots, \min(k, n)$ DO
   Compute $\alpha\left[\lambda, \mathcal{S}_z(U_{N-k}^N)\right]$, by means of recursive Expression (5.29).
   ENDFOR; ENDFOR;
3. FOR $k = 1, \ldots, N$, DO FOR $j = 1, \ldots, \min(k, n)$ DO
   Compute the table of q_{kj} by means of Expression (5.31);
   ENDFOR; ENDFOR;
4. $j = 0$.
5. FOR $k = 1, \ldots, N$ DO
   select unit k with probability $q_{k(n-j)}$.
   IF k is selected THEN $j = j + 1$; ENDIF;
   ENDFOR.

Note that Algorithm 5.8 is weakly sequential because q_{kj} depends on all the units of the population. However, once the q_{kj} are computed, the selection of the sample is immediate. Algorithm 5.8 is the best solution to make simulations because the q_{kj}'s are computed once.

5.6.10 Alternate Method for Computing π from λ

The inclusion probabilities π_k can be derived from q_{kj}. Let

$$r_k = n - \sum_{\ell=1}^{k-1} S_\ell.$$

In the sequential algorithm, r_k is the number of units that remains to be sampled at step k of the algorithm. Obviously $r_1 = n$. Next

$$r_2 = \begin{cases} n-1 & \text{with probability } \pi_k \\ n & \text{with probability } 1 - \pi_k. \end{cases}$$

The inclusion probabilities can be deduced from

$$\pi_k = \sum_{j=1}^{n} q_{kj} \Pr(r_k = j),$$

where the distribution probability of r_k satisfies the recurrence relation:

$$\Pr(r_k = j) = \Pr(r_{k-1} = j)(1 - q_{k-1,j}) + \Pr(r_{k-1} = j+1) q_{k-1,j+1},$$

with $k > 2$, and

$$\Pr(r_1 = j) = \begin{cases} 1 & \text{if } j = n \\ 0 & \text{if } j \neq n. \end{cases}$$

From $\boldsymbol{\lambda}$, it is thus possible to compute $\alpha\left[\boldsymbol{\lambda}, \mathcal{S}_z(U_{N-k}^N)\right]$ by means of recursive Expression (5.29). Next q_{kj} can be computed from $\alpha\left[\boldsymbol{\lambda}, \mathcal{S}_z(U_{N-k}^N)\right]$ by means of Expression (5.31). Finally, $\boldsymbol{\pi}$ can be derived from q_{kj}, which allows defining an alternate procedure of deriving $\boldsymbol{\pi}$ from $\boldsymbol{\lambda}$ in place of Expression (5.13), page 81.

5.6.11 Draw by Draw Procedure for CPS

In order to have a fast implementation of the standard procedure, we use the following result from Chen et al. (1994).

Result 29. *Let $\mathbf{S}$ be a sample selected according to a CPS, A a subset of U and*

$$q_k(A) = \begin{cases} \dfrac{1}{n - \text{card } A} E\left[S_k | S_i = 1 \text{ for all } i \in A\right] & k \notin A \\ 0 & k \in A. \end{cases}$$

Then $q_k(\emptyset) = \pi_k / n$, and

$$q_k(A \cup \{j\}) = \begin{cases} \dfrac{1}{n - \text{card}A - 1} \dfrac{\frac{q_k(A)}{q_j(A)} \exp \lambda_j - \exp \lambda_k}{\exp \lambda_j - \exp \lambda_k} & k \notin A, q_k(A) \neq q_j(A) \\ \dfrac{1 - \sum_{k \notin A, q_k(A) \neq q_j(A)} q_k(A \cup \{j\})}{\text{card}\{k \notin A, q_k(A) = q_j(A)\}} & k \notin A, q_k(A) = q_j(A) \\ 0 & k \in A \cup \{j\}. \end{cases}$$

$$(5.32)$$

Proof. Let $a = \mathrm{card}A$. If $\lambda_j \neq \lambda_k$, we have, for all $j \neq k \in U \backslash A$,

$$(\exp \lambda_j) \sum_{\mathbf{s} \in \mathcal{S}_{n-a}(U \backslash A)} s_k \exp \boldsymbol{\lambda}'\mathbf{s} - (\exp \lambda_k) \sum_{\mathbf{s} \in \mathcal{S}_{n-a}(U \backslash A)} s_j \exp \boldsymbol{\lambda}'\mathbf{s}$$

$$= (\exp \lambda_j) \left(\sum_{\substack{\mathbf{s} \in \mathcal{S}_{n-a}(U \backslash A) \\ s_k=1, s_j=0}} \exp \boldsymbol{\lambda}'\mathbf{s} + \sum_{\substack{\mathbf{s} \in \mathcal{S}_{n-a}(U \backslash A) \\ s_k=1, s_j=1}} \exp \boldsymbol{\lambda}'\mathbf{s} \right)$$

$$- (\exp \lambda_k) \left(\sum_{\substack{\mathbf{s} \in \mathcal{S}_{n-a}(U \backslash A) \\ s_k=0, s_j=j}} \exp \boldsymbol{\lambda}'\mathbf{s} + \sum_{\substack{\mathbf{s} \in \mathcal{S}_{n-a}(U \backslash A) \\ s_k=1, s_j=1}} \exp \boldsymbol{\lambda}'\mathbf{s} \right)$$

$$= (\exp \lambda_j - \exp \lambda_k) \sum_{\mathbf{s} \in \mathcal{S}_{n-a}(U \backslash A)} s_k s_j \exp \boldsymbol{\lambda}'\mathbf{s}. \tag{5.33}$$

Moreover, because

$$q_k(A) = \frac{1}{n-a} \frac{\sum_{\mathbf{s} \in \mathcal{S}_{n-a}(U \backslash A)} s_k \exp \boldsymbol{\lambda}'\mathbf{s}}{\sum_{\mathbf{s} \in \mathcal{S}_{n-a}(U \backslash A)} \exp \boldsymbol{\lambda}'\mathbf{s}}, \quad k \notin A,$$

and

$$\frac{q_k(A)}{q_j(A)} = \frac{\sum_{\mathbf{s} \in \mathcal{S}_{n-a}(U \backslash A)} s_k \exp \boldsymbol{\lambda}'\mathbf{s}}{\sum_{\mathbf{s} \in \mathcal{S}_{n-a}(U \backslash A)} s_j \exp \boldsymbol{\lambda}'\mathbf{s}}, \quad k \notin A,$$

we have

$$\frac{\frac{q_k(A)}{q_j(A)} \exp \lambda_j - \exp \lambda_k}{\exp \lambda_j - \exp \lambda_k}$$

$$= \frac{(\exp \lambda_j) \sum_{\mathbf{s} \in \mathcal{S}_{n-a}(U \backslash A)} s_k \exp \boldsymbol{\lambda}'\mathbf{s} - (\exp \lambda_k) \sum_{\mathbf{s} \in \mathcal{S}_{n-a}(U \backslash A)} s_j \exp \boldsymbol{\lambda}'\mathbf{s}}{(\exp \lambda_j - \exp \lambda_k) \sum_{\mathbf{s} \in \mathcal{S}_{n-a}(U \backslash A)} s_j \exp \boldsymbol{\lambda}'\mathbf{s}}.$$

By (5.33), we obtain

$$\frac{\frac{q_k(A)}{q_j(A)} \exp \lambda_j - \exp \lambda_k}{\exp \lambda_j - \exp \lambda_k}$$

$$= \frac{\sum_{\mathbf{s} \in \mathcal{S}_{n-a}(U \backslash A)} s_k s_j \exp \boldsymbol{\lambda}'\mathbf{s}}{\sum_{\mathbf{s} \in \mathcal{S}_{n-a}(U \backslash A)} s_j \exp \boldsymbol{\lambda}'\mathbf{s}} = \frac{\sum_{\mathbf{s} \in \mathcal{S}_{n-a-1}[U \backslash (A \cup \{j\})]} s_k \exp \boldsymbol{\lambda}'\mathbf{s}}{\sum_{\mathbf{s} \in \mathcal{S}_{n-a-1}[U \backslash (A \cup \{j\})]} \exp \boldsymbol{\lambda}'\mathbf{s}}$$

$$= (n - a - 1) q_k(A \cup \{j\}). \qquad \square$$

The standard draw by draw procedure given in Expression (3.3), page 35, gives Algorithm 5.9. Other procedures that implement the CPS design are proposed in Chen and Liu (1997).

Algorithm 5.9 Draw by draw procedure for CPS

1. Compute $\boldsymbol{\lambda}$ from $\boldsymbol{\pi}$ by means of the Newton method as written in Expression (5.16).
2. Define $A_0 = \emptyset$.
3. FOR $j = 0, \ldots, n-1$, DO
       Select a unit from U with probability $q_k(A_j)$ as defined in Expression (5.32);
       If k_j is the selected unit, THEN $A_{j+1} = A_j \cup \{k_j\}$;
       ENDIF;
     ENDFOR.
4. The selected units are in the set A_n.

5.7 Links Between the Exponential Designs

The following relations can be proved using the definition of conditioning with respect to a sampling design.

- $p_{\text{POISSWR}}(\mathbf{s}, \boldsymbol{\lambda} | \mathcal{R}_n) = p_{\text{MULTI}}(\mathbf{s}, \boldsymbol{\lambda}, n)$,
- $p_{\text{POISSWR}}(\mathbf{s}, \boldsymbol{\lambda} | \mathcal{S}) = p_{\text{POISSWOR}}(\mathbf{s}, \boldsymbol{\lambda})$,
- $p_{\text{POISSWR}}(\mathbf{s}, \boldsymbol{\lambda} | \mathcal{S}_n) = p_{\text{CPS}}(\mathbf{s}, \boldsymbol{\lambda}, n)$,
- $p_{\text{MULTI}}(\mathbf{s}, \boldsymbol{\lambda}, n | \mathcal{S}_n) = p_{\text{CPS}}(\mathbf{s}, \boldsymbol{\lambda}, n)$,
- $p_{\text{POISSWOR}}(\mathbf{s}, \boldsymbol{\lambda} | \mathcal{S}_n) = p_{\text{CPS}}(\mathbf{s}, \boldsymbol{\lambda}, n)$.

These relations are summarized in Figure 5.1 and Table 5.7, page 97.

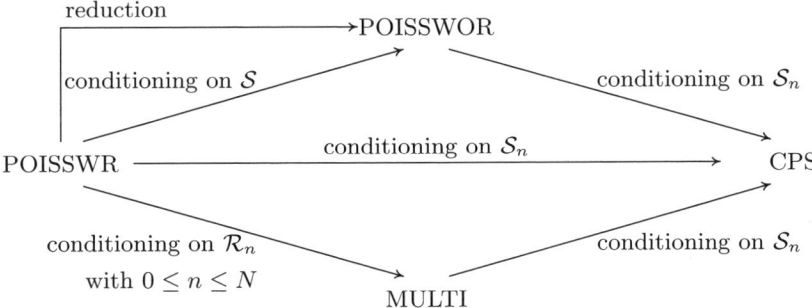

Fig. 5.1. Links between the main exponential sampling designs

5.8 Links Between Exponential Designs and Simple Designs

The following relations are obvious:

- $p_{\text{POISSWR}}(\mathbf{s}, \mathbf{1} \log \theta) = p_{\text{EPPOISSWR}}(\mathbf{s}, \theta)$,
- $p_{\text{MULTI}}(\mathbf{s}, \mathbf{1} \log \theta, n) = p_{\text{SRSWR}}(\mathbf{s}, n)$,

- $p_{\text{POISSWOR}}(\mathbf{s}, \mathbf{1}\log\theta) = p_{\text{BERN}}\left(\mathbf{s}, \pi = \dfrac{\theta}{1+\theta}\right),$
- $p_{\text{CPS}}(\mathbf{s}, \mathbf{1}\log\theta, n) = p_{\text{SRSWOR}}(\mathbf{s}, n).$

5.9 Exponential Procedures in Brewer and Hanif

Brewer and Hanif (1983) have defined a set of properties in order to evaluate the sampling methods. These properties are listed in Table 5.6. Seven exponential procedures are described in Brewer and Hanif (1983); they are presented in Table 5.8, page 98.

Most of them are due to Hájek. Because the way of deriving the inclusion probabilities from the parameter was not yet defined, Hájek proposed different working probabilities to apply the rejective Poisson sampling design. The Hájek procedures do not have probabilities of inclusion strictly proportional to size (not strpps). The Yates and Grundy (1953) procedure and the Carroll and Hartley (1964) methods are very intricate and can be used only for small sample sizes.

Table 5.6. Abbreviations of Brewer and Hanif (1983) and definition of "exact" and "exponential"

strpps	Inclusion probability strictly proportional to size
strwor	Strictly without replacement
n fixed	Number of units in sample fixed
syst	Systematic
d by d	Draw by draw
rej	Rejective
ws	Whole sample
ord	Ordered
unord	Unordered
inexact	Fails to satisfy at least one of the three properties: strpps, strwor, n fixed
n=2 only	Applicable for sample size equal to 2 only
b est var	Estimator of variance generally biased
j p enum	Calculation of joint inclusion probability in sample involves enumeration of all possible selections, or at least a large number of them
j p iter	Calculation of joint inclusion probability in sample involves iteration on computer
not gen app	Not generally applicable
nonrotg	Nonrotating
exact	Satisfies conjointly the three properties: strpps, strwor, n fixed
exponential	Procedure that implements an exponential design

Table 5.7. Main exponential sampling designs

	Notation	POISSWR	MULTI	POISSWOR	CPS
Design	$p(\mathbf{s})$	$\prod_{k\in U}\dfrac{\mu_k^{s_k}e^{-\mu_k}}{s_k!}$	$\dfrac{n!}{n^n}\prod_{k\in U}\dfrac{\mu_k^{s_k}}{s_k!}$	$\prod_{k\in U}\left[\pi_k^{s_k}(1-\pi_k)^{1-s_k}\right]$	$\sum_{s\in\mathcal{S}_n}\exp[\lambda's-\alpha(\lambda,\mathcal{S}_n)]$
Support	$\mathcal{Q}$	$\mathcal{R}$	$\mathcal{R}_n$	$\mathcal{S}$	$\mathcal{S}_n$
Norming constant	$\alpha(\lambda,\mathcal{Q})$	$\sum_{k\in U}\exp\lambda_k$	$\log\dfrac{1}{n!}\left(\sum_{k\in U}\exp\lambda_k\right)^n$	$\log\prod_{k\in U}(1+\exp\lambda_k)$	see Section 5.6.6
Char. function	$\phi(\mathbf{t})$	$\exp\sum_{k\in U}\mu_k(e^{it_k}-1)$	$\left(\dfrac{1}{n}\sum_{k\in U}\mu_k\exp it_k\right)^n$	$\prod_{k\in U}\{1+\pi_k(\exp it_k-1)\}$	not reducible
Replacement	WOR/WR	with repl.	with repl.	without repl.	without repl.
Sample size	$n(\mathbf{S})$	random	fixed	random	fixed
Expectation	μ_k	μ_k	μ_k	π_k	$\pi_k(\lambda,\mathcal{S}_n)$ see Section 5.6.2
Inclusion probability	π_k	$1-e^{-\mu_k}$	$1-(1-\mu_k/n)^n$	π_k	$\pi_k(\lambda,\mathcal{S}_n)$
Variance	Σ_{kk}	μ_k	$\mu_k(n-\mu_k)/n$	π_k^2	see Section 5.6.4
Joint expectation	$\mu_{k\ell},\,k\neq\ell$	μ_k^2	$\mu_k\mu_\ell(n-1)/n$		$\pi_{k\ell}(\lambda,\mathcal{S}_n)$ see Section 5.6.4
Covariance	$\Sigma_{k\ell},\,k\neq\ell$	0	$-\mu_k\mu_\ell/n$	0	see Section 5.6.4
Basic est.	$\widehat{Y}$	$\sum_{k\in U}\dfrac{y_kS_k}{\mu_k}$	$\sum_{k\in U}\dfrac{y_kS_k}{\mu_{k'}}$	$\sum_{k\in U}\dfrac{y_kS_k}{\pi_k}$	$\sum_{k\in U}\dfrac{y_kS_k}{\pi_k}$
Variance	$\mathrm{var}(\widehat{Y})$	$\sum_{k\in U}\dfrac{y_k^2}{\mu_k}$	$\sum_{k\in U}\left(\dfrac{ny_k}{\mu_k}-Y\right)^2$	$\sum_{k\in U}y_k^2\dfrac{1-\pi_k}{\pi_k}$	not reducible
Variance est.	$\widehat{\mathrm{var}}(\widehat{Y})$	$\sum_{k\in U}\dfrac{y_k^2}{\mu_k^2}S_k$	$\dfrac{n}{n-1}\sum_{k\in U}S_k\left(\dfrac{y_k}{\mu_k}-\dfrac{\widehat{Y}_{HH}}{n}\right)^2$	$\sum_{k\in U}S_ky_k^2\dfrac{1-\pi_k}{\pi_k^2}$	not reducible

Table 5.8. Brewer and Hanif list of exponential procedures

Number	Page	Procedure name	strpps	strwor	n fixed	syst	d by d	rej	ws	ord	unord	inexact	n=2 only	b est var	j p enum	j p iter	not gen app	nonrotg	exact	exponential
5	24	Yates-Grundy Rejective Procedure Principal reference: Yates and Grundy (1953) Other ref.: Durbin (1953), Hájek (1964)	+	−	+	−	+	−	+	−	−	−	−	−	+	−	−	−	−	+
14	31	Carroll-Hartley Rejective Procedure Principal reference: Carroll and Hartley (1964) Other ref.: Rao (1963a), Hájek (1964), Rao and Bayless (1969), Bayless and Rao (1970), Cassel et al. (1993, p. 16)	+	+	+	−	−	+	−	−	+	−	−	−	−	+	−	+	+	+
27	39	Poisson Sampling Principal reference: Hájek (1964) Other ref.: Ogus and Clark (1971), Brewer et al. (1972), Brewer et al. (1984), Cassel et al. (1993, p. 17)	+	+	−	+	−	−	−	+	−	−	−	−	−	−	−	−	−	+
28	40	Hájek's "Method I" Principal reference: Hájek (1964)	−	+	−	+	−	+	−	+	−	+	−	−	−	−	−	−	−	+
29	40	Hájek's "Method II" Principal reference: Hájek (1964)	−	+	+	−	−	+	−	+	−	+	−	−	−	−	−	−	−	+
30	40	Hájek's "Method III" Principal reference: Hájek (1964)	−	+	−	+	−	+	−	+	−	+	−	−	−	−	−	−	−	+
31	40	Hájek's "Method IV" Principal reference: Hájek (1964)	−	+	+	−	−	+	−	+	−	+	−	−	−	−	−	−	−	+

6

The Splitting Method

6.1 Introduction

The splitting method, proposed by Deville and Tillé (1998), is a general framework of sampling methods without replacement, with fixed sample size and unequal probabilities. The basic idea consists of splitting the inclusion probability into two or several new vectors. Next, one of these vectors is selected randomly in such a way that the average of the vectors is the vector of inclusion probabilities. This simple step is repeated until a sample is obtained. The splitting method is thus a martingale algorithm. It includes all the draw by draw and sequential procedures and it allows deriving a large number of unequal probability methods. Moreover, many well-known procedures of unequal probabilities can be formulated under the form of a splitting. The presentation can thus be standardized, which allows a simpler comparison of the procedures. In this chapter, we present the fundamental splitting algorithm. Two variants can be defined depending on whether a vector or a direction is chosen. Several applications are given: minimum support design, splitting into simple random sampling, pivotal procedure, and the generalized Sunter procedure. Next, a general method of splitting into several vectors is presented. The particular cases are the Brewer procedure, the eliminatory method, the Tillé elimination method, the generalized Midzuno procedure, and the Chao procedure.

6.2 Splitting into Two Vectors

6.2.1 A General Technique of Splitting into Two Vectors

Consider a vector of inclusion probabilities $\boldsymbol{\pi} = (\pi_1 \; \cdots \; \pi_N)'$ such that

$$n(\boldsymbol{\pi}) = \sum_{k \in U} \pi_k = n,$$

and n is an integer. The general splitting technique into two vectors is presented in Algorithm 6.1. As shown in Figure 6.1, at each step $t = 0, 1, 2, \dots$, vector $\boldsymbol{\pi}(t)$ is split into two other vectors. The splitting method is based on a random transformation of vector $\boldsymbol{\pi}$ until a sample is obtained.

Algorithm 6.1 General splitting method into two vectors

---

1. INITIALIZE $\boldsymbol{\pi}(0) = \boldsymbol{\pi}$, and $t = 0$.
2. WHILE $\boldsymbol{\pi}(t) \notin \{0, 1\}^N$ DO
    - Construct any pair of vectors $\boldsymbol{\pi}^a(t)$, $\boldsymbol{\pi}^b(t) \in \mathbb{R}^N$, and a scalar $\alpha(t) \in [0, 1]$ such that
      $$\alpha(t)\boldsymbol{\pi}^a(t) + [1 - \alpha(t)]\,\boldsymbol{\pi}^b(t) = \boldsymbol{\pi}(t),$$
      $$n\left[\boldsymbol{\pi}^a(t)\right] = n\left[\boldsymbol{\pi}^b(t)\right] = n\left[\boldsymbol{\pi}(t)\right],$$
      $$0 \le \pi_k^a(t) \le 1, \text{ and } 0 \le \pi_k^b(t) \le 1;$$
    - $\boldsymbol{\pi}(t + 1) = \begin{cases} \boldsymbol{\pi}^a(t) \text{ with probability } \alpha(t) \\ \boldsymbol{\pi}^b(t) \text{ with probability } 1 - \alpha(t); \end{cases}$
    - $t = t + 1$;
    ENDWHILE.
3. The selected sample is $\mathbf{S} = \boldsymbol{\pi}(t)$.

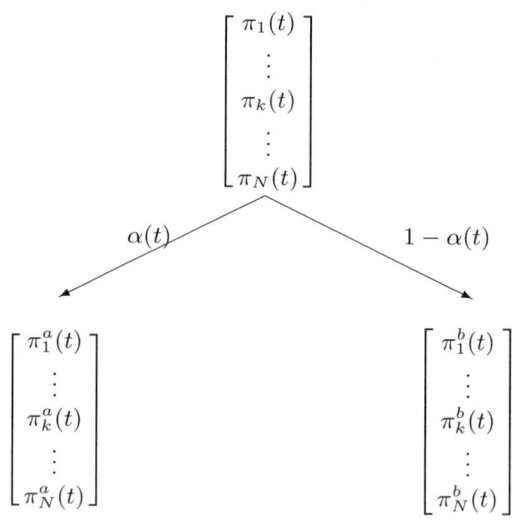

Fig. 6.1. Splitting into two vectors

Vector $\boldsymbol{\pi}(t)$ is a martingale because

$$E\left[\boldsymbol{\pi}(t)|\boldsymbol{\pi}(t-1),\ldots,\boldsymbol{\pi}(1)\right] = \boldsymbol{\pi}(t-1).$$

It follows that

$$E\left[\boldsymbol{\pi}(t)\right] = \boldsymbol{\pi},$$

and thus the splitting method selects a sample with the inclusion probabilities equal to $\boldsymbol{\pi}$. At each step of the algorithm, the problem is reduced to another sampling problem with unequal probabilities. If the splitting is such that one or several of the π_k^a and the π_k^b are equal to 0 or 1, the sampling problem will be simpler at the next step. Indeed, once a component of $\boldsymbol{\pi}(t)$ becomes an integer, it must remain an integer for all the following steps of the method. The splitting method allows defining a large family of algorithms that includes many known procedures.

6.2.2 Splitting Based on the Choice of $\boldsymbol{\pi}^a(t)$

A first way to construct $\boldsymbol{\pi}^a(t)$, $\boldsymbol{\pi}^b(t)$, and $\alpha(t)$ consists of arbitrarily fixing $\boldsymbol{\pi}^a(t)$ and then of deriving appropriate values for $\boldsymbol{\pi}^b(t)$ and $\alpha(t)$, which gives Algorithm 6.2.

Algorithm 6.2 Splitting method based on the choice of $\boldsymbol{\pi}^a(t)$

INITIALIZE $\boldsymbol{\pi}(0) = \boldsymbol{\pi}$;

FOR $t = 0, 1, 2\ldots$, and until a sample is obtained, DO

1. Generate any vector of probabilities $\boldsymbol{\pi}^a(t)$ such that $n[\boldsymbol{\pi}^a(t)] = n(\boldsymbol{\pi})$ and $\pi_k^a(t) = \pi_k(t)$ if $\pi_k(t) \in \{0,1\}$. The vector $\boldsymbol{\pi}^a(t)$ can be generated randomly or not;

2. Define $\alpha(t)$ as the largest real number in $[0,1]$ that satisfies

$$0 \le \frac{\pi_k(t) - \alpha(t)\pi_k^a(t)}{1 - \alpha(t)} \le 1, \text{ for all } k \in U;$$

that is,

$$\alpha(t) = \min\left[\min_{\substack{k \in U \\ 0 < \pi_k(t) < 1}} \frac{\pi_k(t)}{\pi_k^a(t)};\ \min_{\substack{k \in U \\ 0 < \pi_k(t) < 1}} \frac{1 - \pi_k(t)}{1 - \pi_k^a(t)}\right]; \tag{6.1}$$

3. Define

$$\boldsymbol{\pi}^b = \frac{\boldsymbol{\pi} - \alpha(t)\boldsymbol{\pi}^a(t)}{1 - \alpha(t)}; \tag{6.2}$$

4. Select $\boldsymbol{\pi}(t+1) = \begin{cases} \boldsymbol{\pi}^a & \text{with probability } \alpha(t) \\ \boldsymbol{\pi}^b & \text{with probability } 1 - \alpha(t); \end{cases}$

ENDFOR.

6.2.3 Methods Based on the Choice of a Direction

A second way to construct $\boldsymbol{\pi}^a(t)$, $\boldsymbol{\pi}^b(t)$, and $\alpha(t)$ consists of arbitrarily choosing at each step a direction $\mathbf{u}(t)$ used to determine the splitting, which gives Algorithm 6.3.

Algorithm 6.3 Splitting method based on the choice of a direction

INITIALIZE $\boldsymbol{\pi}(0) = \boldsymbol{\pi}$;

FOR $t = 0, 1, 2 \ldots$, and until a sample is obtained, DO

1. Define $U(t) = \{k \in U \mid 0 < \pi_k(t) < 1\}$;
2. Choose arbitrarily (randomly or not) a nonnull vector $\mathbf{u}(t) = (u_k(t)) \in \mathbb{R}^N$ such that

$$u_k(t) = 0 \quad \text{for all } k \notin U(t), \quad \text{and} \quad \sum_{k \in U} u_k(t) = 0;$$

3. Compute $\lambda_1^*(t)$ and $\lambda_2^*(t)$, the largest positive values of $\lambda_1(t)$ and $\lambda_2(t)$, such that

$$0 \le \pi_k(t) + \lambda_1(t)u_k(t) \le 1, \text{ and } 0 \le \pi_k(t) - \lambda_2(t)u_k(t) \le 1;$$

for all $k \in U$.

4. Define

$$\boldsymbol{\pi}^a = \boldsymbol{\pi}(t) + \lambda_1^*(t)\mathbf{u}(t),$$

$$\boldsymbol{\pi}^b = \boldsymbol{\pi}(t) - \lambda_2^*(t)\mathbf{u}(t),$$

$$\alpha(t) = \frac{\lambda_2^*(t)}{\lambda_1^*(t) + \lambda_2^*(t)};$$

5. Select

$$\boldsymbol{\pi}(t+1) = \begin{cases} \boldsymbol{\pi}^a & \text{with probability } \alpha(t) \\ \boldsymbol{\pi}^b & \text{with probability } 1 - \alpha(t); \end{cases}$$

ENDFOR.

6.2.4 Minimum Support Design

Wynn (1977, Theorem 1) proved that it is always possible to define a sampling design with any fixed first-order inclusion probabilities by using only N samples $\mathbf{s}$ such that $p(\mathbf{s}) > 0$. This sampling design is called the minimum support design. However Wynn's result is not constructive. The minimum support design was already proposed by Jessen (1969) in a more restrictive context (see also Brewer and Hanif, 1983, procedures 35 and 36, pp. 42-43). Hedayat et al. (1989, Theorem 2) have also proposed a related procedure in the context of a method of emptying boxes, but its implementation is limited to inclusion probabilities that can be written as rational numbers. The minimum support design is presented in Algorithm 6.4.

With Algorithm 6.4, if the $\pi_k^a(t)$ are selected, the sample is automatically selected. If the $\pi_k^b(t)$ are selected, the problem is reduced to the selection of a sample from a smaller population. In at most N steps, the sample is selected.

Algorithm 6.4 Minimum support procedure

INITIALIZE $\boldsymbol{\pi}(0) = \boldsymbol{\pi}$;

FOR $t = 0, 1, 2 \ldots$, and until a sample is obtained, DO

1. Define
   $A_t = \{k | \pi_k(t) = 0\}$, $B_t = \{k | \pi_k(t) = 1\}$, and $C_t = \{k | 0 < \pi_k(t) < 1\}$;
2. Select a subset D_t of C_t such that card $D_t = n -$ card B_t (D_t can be selected randomly or not);
3. Define

$$\pi_k^a(t) = \begin{cases} 0 & \text{if } k \in A_t \cup (C_t \backslash D_t) \\ 1 & \text{if } k \in B_t \cup D_t, \end{cases}$$

$$\alpha(t) = \min\{1 - \max_{k \in (C_t \backslash D_t)} \pi_k, \min_{k \in D_t} \pi_k\},$$

and

$$\pi_k^b = \begin{cases} 0 & \text{if } k \in A_t \\ 1 & \text{if } k \in B_t \\ \dfrac{\pi_k(t)}{1 - \alpha(t)} & \text{if } k \in (C_t \backslash D_t) \\ \dfrac{\pi_k(t) - \alpha(t)}{1 - \alpha(t)} & \text{if } k \in D_t; \end{cases}$$

4. Select $\boldsymbol{\pi}(t+1) = \begin{cases} \boldsymbol{\pi}^a & \text{with probability } \alpha(t) \\ \boldsymbol{\pi}^b & \text{with probability } 1 - \alpha(t); \end{cases}$

ENDFOR.

Example 12. Suppose that $N = 6, n = 3$, $\boldsymbol{\pi} = (0.07\ 0.17\ 0.41\ 0.61\ 0.83\ 0.91)'$. At each step, the subset D_t consists of the largest noninteger values of $\boldsymbol{\pi}(t)$. In this case, the splitting is completed in four steps. The vector of inclusion probabilities is split into two parts, as given in columns 2 and 3 of Table 6.1.

Table 6.1. Minimum support design for Example 12

π_k	Step 1 $\alpha(0) = 0.59$		Step 2 $\alpha(1) = 0.585$		Step 3 $\alpha(2) = 0.471$		Step 4 $\alpha(3) = 0.778$	
$\boldsymbol{\pi}(0)$	$\boldsymbol{\pi}^a(0)$	$\boldsymbol{\pi}^b(0)$	$\boldsymbol{\pi}^a(1)$	$\boldsymbol{\pi}^b(1)$	$\boldsymbol{\pi}^a(2)$	$\boldsymbol{\pi}^b(2)$	$\boldsymbol{\pi}^a(3)$	$\boldsymbol{\pi}^b(3)$
0.07	0	0.171	0	0.412	0	0.778	1	0
0.17	0	0.415	0	1	1	1	1	1
0.41	0	1	1	1	1	1	1	1
0.61	1	0.049	0	0.118	0	0.222	0	1
0.83	1	0.585	1	0	0	0	0	0
0.91	1	0.780	1	0.471	1	0	0	0

With probability $\alpha(0) = 0.59$, the sample $(0, 0, 0, 1, 1, 1)$ is selected and, with probability $1 - \alpha(0) = 0.41$, another sampling is applied with unequal probabilities given by $(0.171, 0.415, 1, 0.049, 0.585, 0.780)'$. At step 2, the splitting is applied again to this vector and, in four steps, the sample is selected as shown in the splitting tree presented in Figure 6.2. The sampling design is thus given by:

$$p((0, 0, 0, 1, 1, 1)') = 0.59,$$
$$p((0, 0, 1, 0, 1, 1)') = (1 - 0.59) \times 0.585 = 0.24,$$
$$p((0, 1, 1, 0, 0, 1)') = (1 - 0.59 - 0.24) \times 0.471 = 0.08,$$
$$p((1, 1, 1, 0, 0, 0)') = (1 - 0.59 - 0.24 - 0.08) \times 0.778 = 0.07,$$
$$p((0, 1, 1, 1, 0, 0)') = (1 - 0.59 - 0.24 - 0.08 - 0.7) = 0.02.$$

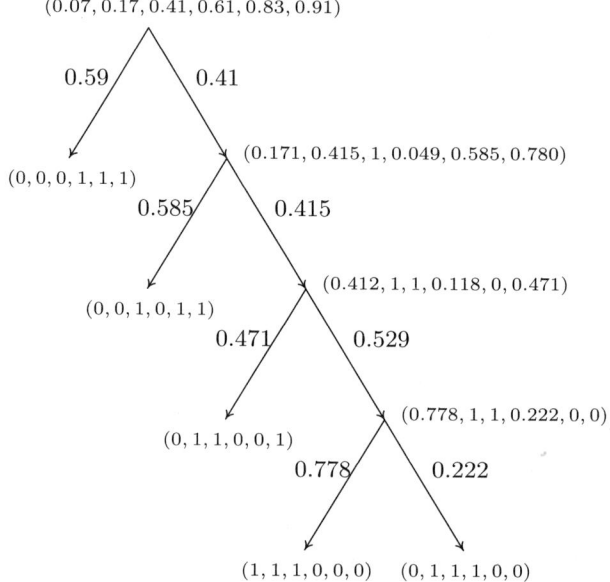

Fig. 6.2. Splitting tree for the minimum support design for Example 12

6.2.5 Splitting into Simple Random Sampling

This method also splits the inclusion probabilities into two parts. One part is a vector with equal probabilities, so a simple random sample can be selected. Algorithm 6.5 is an implementation of this splitting method into simple random sampling.

Example 13. With the same π as in Example 12; that is, $N = 6, n = 3$, $\pi = (0.07 \ 0.17 \ 0.41 \ 0.61 \ 0.83 \ 0.91)'$, the result of the method is given in Table 6.2 and the splitting tree is presented in Figure 6.3.

Algorithm 6.5 Splitting procedure into simple random sampling

INITIALIZE $\boldsymbol{\pi}(0) = \boldsymbol{\pi}$.

FOR $t = 0, 1, 2 \ldots$, and until a sample is obtained, DO

1. Define $A_t = \{k | \pi_k(t) = 0\}$, and $B_t = \{k | \pi_k(t) = 1\}$;
2. Compute

$$
\pi_k^a(t) = \begin{cases} \dfrac{n - \mathrm{card} B_t}{N - \mathrm{card} A_t - \mathrm{card} B_t} & \text{if } k \notin A_t \cup B_t \\ 0 & \text{if } k \in A_t \\ 1 & \text{if } k \in B_t, \end{cases}
$$

$$
\alpha(t) = \min \left\{ \frac{N - \mathrm{card} A_t - \mathrm{card} B_t}{n - \mathrm{card} B_t} \min_{k \in U \backslash (A \cup B)} \pi_k(t); \right.
$$
$$
\left. \frac{N - \mathrm{card} A_t - \mathrm{card} B_t}{N - n - \mathrm{card} B_t} \left[1 - \max_{k \in U \backslash (A \cup B)} \pi_k(t) \right] \right\},
$$

and

$$
\boldsymbol{\pi}^b(t) = \frac{\boldsymbol{\pi}(t) - \alpha(t) \boldsymbol{\pi}^b}{1 - \alpha(t)};
$$

3. Generate u, a uniform random variable in [0,1];
4. IF $\alpha(t) < u$ THEN
       select a sample $\mathbf{s}$ by means of SRS with inclusion probabilities $\boldsymbol{\pi}^a(t)$;
       STOP;
   ELSE
       IF $\boldsymbol{\pi}^b(t)$ is a sample, select $\mathbf{s} = \boldsymbol{\pi}^b(t)$ and STOP;
       ELSE $\boldsymbol{\pi}(t+1) = \boldsymbol{\pi}^b(t)$;
       ENDIF;
   ENDIF;

ENDFOR.

Table 6.2. Splitting into simple random sampling for Example 13

π_k	Step 1 $\alpha(0) = 0.14$		Step 2 $\alpha(1) = 0.058$		Step 3 $\alpha(2) = 0.173$		Step 4 $\alpha(3) = 0.045$		Step 5 $\alpha(4) = 0.688$	
$\boldsymbol{\pi}(0)$	$\boldsymbol{\pi}^a(0)$	$\boldsymbol{\pi}^b(0)$	$\boldsymbol{\pi}^a(1)$	$\boldsymbol{\pi}^b(1)$	$\boldsymbol{\pi}^a(2)$	$\boldsymbol{\pi}^b(2)$	$\boldsymbol{\pi}^a(3)$	$\boldsymbol{\pi}^b(3)$	$\boldsymbol{\pi}^a(4)$	$\boldsymbol{\pi}^b(4)$
0.07	0.5	0	0	0	0	0	0	0	0	0
0.17	0.5	0.116	0.6	0.086	0.5	0	0	0	0	0
0.41	0.5	0.395	0.6	0.383	0.5	0.358	0.667	0.344	0.5	0
0.61	0.5	0.628	0.6	0.630	0.5	0.657	0.667	0.656	0.5	1
0.83	0.5	0.884	0.6	0.901	0.5	0.985	0.667	1	1	1
0.91	0.5	0.977	0.6	1	1	1	1	1	1	1

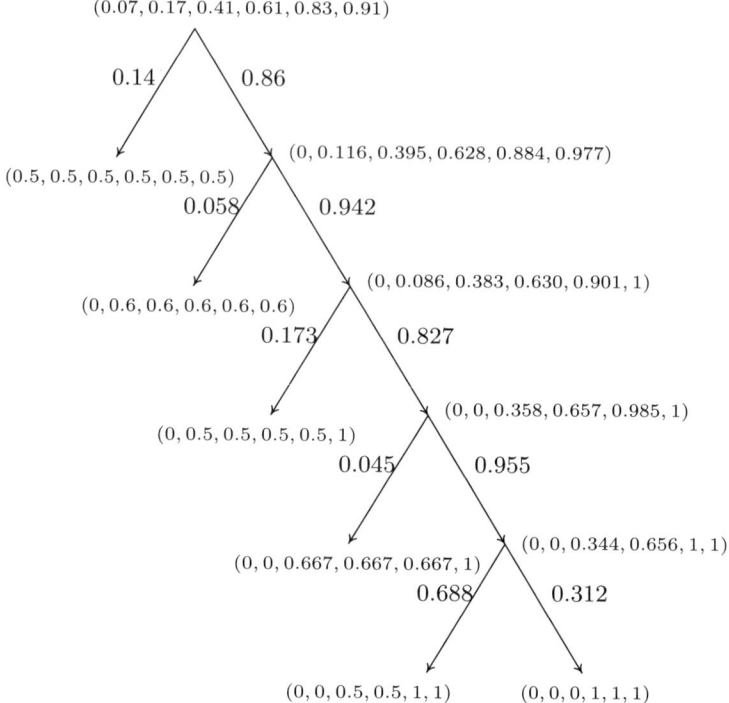

Fig. 6.3. Splitting tree for the splitting method into simple random sampling of Example 13

6.2.6 The Pivotal Method

The pivotal method proposed by Deville and Tillé (1998) consists of splitting the vector of inclusion probabilities into two parts, but, at each step, only two inclusion probabilities are modified. The basic step is as follows. At step t, select two units from the population denoted i and j such that $0 < \pi_i(t) < 1$ and $0 < \pi_j(t) < 1$.

If $\pi_i(t) + \pi_j(t) > 1$, then

$$\alpha(t) = \frac{1 - \pi_j(t)}{2 - \pi_i(t) - \pi_j(t)},$$

$$\pi_k^a(t) = \begin{cases} \pi_k(t) & k \in U \backslash \{i, j\} \\ 1 & k = i \\ \pi_i(t) + \pi_j(t) - 1 & k = j, \end{cases}$$

$$\pi_k^b(t) = \begin{cases} \pi_k(t) & k \in U \backslash \{i, j\} \\ \pi_i(t) + \pi_j(t) - 1 & k = i \\ 1 & k = j. \end{cases}$$

On the other hand, if $\pi_i(t) + \pi_j(t) < 1$, then

$$\alpha = \frac{\pi_i(t)}{\pi_i(t) + \pi_j(t)},$$

$$\pi_k^a(t) = \begin{cases} \pi_k(t) & k \in U \backslash \{i,j\} \\ \pi_i(t) + \pi_j(t) & k = i \\ 0 & k = j, \end{cases}$$

$$\pi_k^b(t) = \begin{cases} \pi_k(t) & k \in U \backslash \{i,j\} \\ 0 & k = i \\ \pi_i(t) + \pi_j(t) & k = j. \end{cases}$$

In the first case, a one is allocated to only one inclusion probability. In the second case, a zero is allocated to only one inclusion probability. The problem is thus reduced to a population of size $N - 1$. In at most N steps, a solution is obtained. Algorithm 6.6 is a quick and strictly sequential implementation of the pivotal method.

Algorithm 6.6 Pivotal procedure

1. DEFINITION a, b, u REAL; i, j, k INTEGER;
2. $a = \pi_1; b = \pi_2; i = 1; j = 2;$
3. FOR $k = 1, \ldots, N$ DO $s_k = 0$; ENDFOR
4. $k = 3;$
5. WHILE $k \leq N$ DO
   u = uniform random variable in [0,1];
   IF $a + b > 1$ THEN | IF $u < \frac{1-b}{2-a-b}$ THEN $b = a + b - 1; a = 1;$
   | ELSE $a = a + b - 1; b = 1;$
   | ENDIF;
   ELSE | IF $u < \frac{b}{a+b}$ THEN $b = a + b; a = 0;$
   | ELSE $a = a + b; b = 0;$
   | ENDIF;
   ENDIF;
   IF a is integer and $k \leq N$ THEN $s_i = a; a = \pi_k; i = k; k = k + 1$; ENDIF;
   IF b is integer and $k \leq N$ THEN $s_j = b; b = \pi_k; j = k; k = k + 1$; ENDIF;
   ENDWHILE;
6. u = uniform random variable in [0,1];
   IF $a + b¿ 1$ THEN | IF $u < \frac{1-b}{2-a-b}$ THEN $b = a + b - 1; a = 1;$
   | ELSE $a = a + b - 1; b = 1;$
   | ENDIF;
   ELSE | IF $u < \frac{b}{a+b}$ THEN $b = a + b; a = 0;$
   | ELSE $a = a + b; b = 0;$
   | ENDIF;
   ENDIF;
   IF a is integer and $k \leq N$ THEN $s_i = a; a = \pi_k; i = k; k = k + 1$; ENDIF;
   IF b is integer and $k \leq N$ THEN $s_j = b; b = \pi_k; j = k; k = k + 1$; ENDIF;
   $s_i = a; s_j = b.$

If the units are taken in a fixed order, most of the joint inclusion probabilities are equal to zero. In order to overcome this problem, the procedure
can be randomized, as with systematic sampling (see Section 7.1, page 124),
by randomly sorting the population before applying the method.

6.2.7 Random Direction Method

In the random direction method, vectors $\boldsymbol{\pi}^a$ and $\boldsymbol{\pi}^b$ are constructed randomly.
A random direction is chosen by means of a vector $\mathbf{u}$ with zero mean, which
gives Algorithm 6.7.

Algorithm 6.7 Random direction procedure

FOR $t = 0, 1, 2 \ldots$, and until a sample is obtained, DO

1. Define $U(t) = \{k \in U | 0 < \pi_k(t) < 1\}$;

2. Generate a random vector $\mathbf{v}(t)$, where $v_k(t) = \begin{cases} N(0,1) & \text{if } k \in U(t) \\ 0 & \text{if } k \notin U(t), \end{cases}$ where
   $N(0,1)$ is a standard normal distribution.

3. Define $\mathbf{u}(t) = [u_k(t)]$, where

$$u_k(t) = \begin{cases} v_k(t) - \dfrac{1}{\operatorname{card} U(t)} \displaystyle\sum_{k \in U(t)} v_k(t) & \text{if } k \in U(t) \\ 0 & \text{if } k \notin U(t); \end{cases}$$

4. Compute $\lambda_1^*(t)$ and $\lambda_2^*(t)$, the largest values of $\lambda_1(t)$ and $\lambda_2(t)$, such that

$$0 \le \boldsymbol{\pi}(t) + \lambda_1(t)\mathbf{u}(t) \le 1, \text{ and } 0 \le \boldsymbol{\pi}(t) - \lambda_2(t)\mathbf{u}(t) \le 1;$$

5. $\boldsymbol{\pi}^a = \boldsymbol{\pi}(t) + \lambda_1^*(t)\mathbf{u}(t), \quad \boldsymbol{\pi}^b = \boldsymbol{\pi}(t) - \lambda_2^*(t)\mathbf{u}(t), \quad \alpha(t) = \dfrac{\lambda_2^*(t)}{[\lambda_1^*(t) + \lambda_2^*(t)]}$;

6. Select $\boldsymbol{\pi}(t+1) = \begin{cases} \boldsymbol{\pi}^a & \text{with probability } \alpha(t) \\ \boldsymbol{\pi}^b & \text{with probability } 1 - \alpha(t); \end{cases}$

ENDFOR.

6.2.8 Generalized Sunter Method

Sunter (1977, 1986) has proposed a sequential procedure that is not generally
applicable to any unequal probability vector of inclusion probabilities. The
Sunter procedure only works when the units are sorted in decreasing order
and when the smallest units have equal inclusion probabilities. In the framework of the splitting procedure, Deville and Tillé (1998) have generalized the
Sunter method in such a way that it is applicable to any vector of inclusion
probabilities. Algorithm 6.8 is an implementation of the generalized Sunter
procedure and is a particular case of the method based on the choice of $\boldsymbol{\pi}^a(t)$
(see Algorithm 6.2, page 101).

Algorithm 6.8 Generalized Sunter procedure

1. INITIALIZE $\boldsymbol{\pi}(1) = \boldsymbol{\pi}$;
2. FOR $t = 1, 2 \ldots, N$ DO
   IF $\pi_t(t)$ is an integer THEN
   $\boldsymbol{\pi}(t+1) = \boldsymbol{\pi}(t)$;
   ELSE
   Define $A(t) = \{k \in U | \pi_k(t) \in \{0,1\}\}$ and $B(t) = \{k \in U | 0 < \pi_k(t) < 1\}$;

$$\pi_k^a(t) = \begin{cases} \pi_k(t) & \text{if } k \in A(t) \\ 1 & \text{if } k = t \\ \pi_k(t)\dfrac{\sum_{k \in B(t)} \pi_k(t) - 1}{\sum_{k \in B(t)} \pi_k(t) - \pi_t(t)} & \text{if } k \in B(t), k \neq t; \end{cases}$$

   Define $\alpha(t)$ as the largest real number that satisfies

$$0 \le \frac{\pi_k(t) - \alpha(t)\pi_k^a(t)}{1 - \alpha(t)} \le 1, \text{ for all } k \in U;$$

that is,

$$\alpha(t) = \min \left[\min_{\substack{k \in U \\ 0 < \pi_k(t) < 1}} \frac{\pi_k(t)}{\pi_k^a(t)}; \min_{\substack{k \in U \\ 0 < \pi_k(t) < 1}} \frac{1 - \pi_k(t)}{1 - \pi_k^a(t)} \right]; \qquad (6.3)$$

   Define

$$\boldsymbol{\pi}^b = \frac{\boldsymbol{\pi}_k - \alpha(t)\boldsymbol{\pi}^a(t)}{1 - \alpha(t)}; \qquad (6.4)$$

   Select $\boldsymbol{\pi}(t+1) = \begin{cases} \boldsymbol{\pi}^a & \text{with probability } \alpha(t) \\ \boldsymbol{\pi}^b & \text{with probability } 1 - \alpha(t); \end{cases}$
   ENDIF;
   ENDFOR;
3. IF $\boldsymbol{\pi}(N)$ is not a sample, THEN $\boldsymbol{\pi}(1) = \boldsymbol{\pi}(N)$ and GOTO STEP 2; ENDIF.

It is interesting to analyze the first step of the generalized Sunter procedure. First, define

$$\pi_k^a(1) = \begin{cases} 1 & \text{if } k = 1 \\ \pi_k(1)\frac{n-1}{n-\pi_1(1)} & \text{if } k \neq 1. \end{cases}$$

Next, define $\alpha(1)$ and $\boldsymbol{\pi}^b(1)$ by the usual method as given in Expressions (6.3) and (6.4):

$$\alpha(1) = \min \left[\min_{\substack{k \in U \\ 0 < \pi_k(1) < 1}} \frac{\pi_k(1)}{\pi_k^a(1)}; \min_{\substack{k \in U \\ 0 < \pi_k(1) < 1}} \frac{1 - \pi_k(1)}{1 - \pi_k^a(1)} \right]$$

$$= \min \left[\pi_1(1); \frac{[n - \pi_1(1)][1 - \pi_m(1)]}{n[1 - \pi_m(1)] + \pi_m(1) - \pi_1(1)} \right],$$

where $\pi_m(1)$ is the largest $\pi_k(1)$, for $k \in U \backslash \{1\}$, which gives, after some algebra:

$$\alpha(1) = \begin{cases} \pi_1(1) & \text{if } \dfrac{n\pi_m(1)}{n - \pi_1(1)} \leq 1 \\[3mm] \dfrac{[n - \pi_1(1)][1 - \pi_m(1)]}{n[1 - \pi_m(1)] + \pi_m(1) - \pi_1(1)} & \text{if } \dfrac{n\pi_m(1)}{n - \pi_1(1)} > 1. \end{cases}$$

Finally,

$$\pi^b(1) = \frac{\pi_k(1) - \alpha(1)\pi^a(1)}{1 - \alpha(1)}.$$

Two cases can be distinguished.

Case 1. If $n\pi_m(1)/[n - \pi_1(1)] \leq 1$, select

$$\pi(2) = \begin{cases} \pi^a(1) = [\pi_k^a(1)] & \text{with probability } \pi_1(1) \\ \pi^b(1) = [\pi_k^b(1)] & \text{with probability } 1 - \pi_1(1), \end{cases}$$

where

$$\pi_k^a(1) = \begin{cases} 1 & k = 1 \\ \pi_k(1)\dfrac{n-1}{n - \pi_1(1)} & k \neq 1, \end{cases}$$

and

$$\pi_k^b(1) = \begin{cases} 0 & k = 1 \\ \pi_k(1)\dfrac{n}{n - \pi_1(1)} & k \neq 1. \end{cases}$$

Case 2. If $n\pi_m(1)/[n - \pi_1(1)] > 1$, select

$$\pi(2) = \begin{cases} \pi^a(1) = [\pi_k^a(1)] & \text{with probability } \alpha(1) \\ \pi^b(1) = [\pi_k^b(1)] & \text{with probability } 1 - \alpha(1), \end{cases}$$

where

$$\alpha(1) = \frac{[n - \pi_1(1)][1 - \pi_m(1)]}{n[1 - \pi_m(1)] + \pi_m(1) - \pi_1(1)},$$

$$\pi_k^a(1) = \begin{cases} 1 & k = 1 \\ \pi_k(1)\dfrac{n-1}{n - \pi_1(1)} & k \neq 1, \end{cases}$$

and

$$\pi_k^b(1) = \begin{cases} \pi_1(1) - [1 - \pi_m(1)][n - \pi_1(1)]/\pi_m(1) & k = 1 \\ \pi_k(1)/\pi_m(1) & k \neq 1. \end{cases}$$

The basic Sunter (1977, 1986) method considers Case 1 only. For this reason, the Sunter method is not generally applicable to any vector of inclusion probabilities. After the first step, the same elementary procedure is applied to the noninteger units of vector $\pi(2)$ and so on.

Example 14. With the same data as in Example 12; that is, $N = 6, n = 3$, $\pi = (0.07\ 0.17\ 0.41\ 0.61\ 0.83\ 0.91)'$, the first step of Sunter's method is presented in Table 6.3.

Table 6.3. Generalized Sunter method for Example 14

π_k	$\alpha = 0.07$ π_k^a	$1 - \alpha = 0.93$ π_k^b
0.07	1	0
0.17	0.12	0.17
0.41	0.28	0.42
0.61	0.42	0.62
0.83	0.57	0.85
0.91	0.62	0.93

6.3 Splitting into M Vectors

6.3.1 A General Method of Splitting into M Vectors

The splitting procedure can be generalized to a splitting technique into M vectors of inclusion probabilities, as presented in Algorithm 6.9.

Algorithm 6.9 General splitting method into M vectors

---

1. INITIALIZE $\boldsymbol{\pi}(0) = \boldsymbol{\pi}$, and $t = 0$
2. WHILE $\boldsymbol{\pi}(t) \notin \{0,1\}^N$ DO
   a) Construct a set of M vectors $\boldsymbol{\pi}^{(i)}(t), \in \mathbb{R}^N, i = 1, \ldots, M$, and a set of M scalars $\alpha_i(t) \in [0,1]$, such that

$$\begin{cases} \sum_{i=1}^{M} \alpha_i(t)\boldsymbol{\pi}^{(i)} = \boldsymbol{\pi}(t), \\ n\left[\boldsymbol{\pi}^{(i)}(t)\right] = n\left[\boldsymbol{\pi}(t)\right], \text{ for all } i = 1, \ldots, M, \\ 0 \leq \pi_k^{(i)}(t) \leq 1, \text{ for all } i = 1, \ldots, M; \end{cases}$$

   b) $\boldsymbol{\pi}(t+1) = \begin{cases} \boldsymbol{\pi}^{(1)}, & \text{with probability } \alpha_1(t) \\ \vdots \\ \boldsymbol{\pi}^{(i)}, & \text{with probability } \alpha_i(t) \\ \vdots \\ \boldsymbol{\pi}^{(M)}, & \text{with probability } \alpha_M(t); \end{cases}$

   c) $t = t + 1$;
   ENDWHILE;
3. The selected sample is $\mathbf{S} = \boldsymbol{\pi}(t)$.

Note that $\pi_k^{(i)}(t) = \pi_k(t)$, if $\pi_k(t) = \{0,1\}$, for all $i = 1, \ldots, M$. Figure 6.4 shows the basic step of splitting into M vectors.

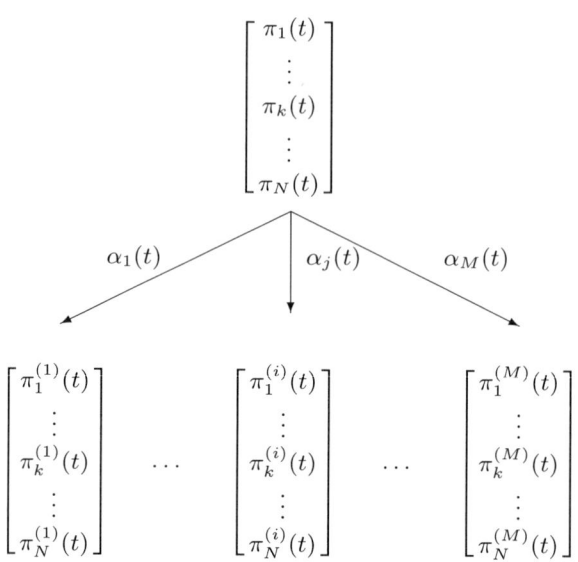

Fig. 6.4. Splitting into M vectors

6.3.2 Brewer's Method

Brewer's method was first proposed for the particular case $n = 2$ in Brewer (1963a) and was discussed in Rao and Bayless (1969); Rao and Singh (1973); Sadasivan and Sharma (1974), and Cassel et al. (1993). Next, Brewer (1975) generalized this method for any sample size (see also Brewer and Hanif, 1983, procedure 8, p. 26). This method is a draw by draw procedure. Thus, in n steps, the sample is selected, but a presentation in a splitting form is more understandable.

For simplicity, we present only the first step of the method. First, define

$$
\alpha_j = \left\{ \sum_{z=1}^{N} \frac{\pi_z(n - \pi_z)}{1 - \pi_z} \right\}^{-1} \frac{\pi_j(n - \pi_j)}{1 - \pi_j}.
$$

Next, compute

$$
\pi_k^{(j)} =
\begin{cases}
\dfrac{\pi_k(n - 1)}{n - \pi_j} & \text{if } k \neq j \\
1 & \text{if } k = j.
\end{cases}
$$

The validity of Brewer's method comes from the following result.

Result 30. *For Brewer's method,*

$$\sum_{j=1}^{N} \alpha_j \pi_k^{(j)} = \pi_k,$$

for all $k = 1, \ldots, N$.

Proof. If we denote

$$C = \left[\sum_{z=1}^{N} \frac{\pi_z(n - \pi_z)}{1 - \pi_z} \right]^{-1},$$

we obtain

$$\sum_{j=1}^{N} \alpha_j \pi_k^{(j)} = \sum_{\substack{j=1 \\ j \neq k}}^{N} C \frac{\pi_k \pi_j(n - 1)}{1 - \pi_j} + C \frac{\pi_k(n - \pi_k)}{1 - \pi_k}$$

$$= C\pi_k \left\{ \sum_{j=1}^{N} \frac{\pi_j(n - 1)}{1 - \pi_j} + n \right\} = C\pi_k \sum_{j=1}^{N} \left[\frac{\pi_j(n - 1)}{1 - \pi_j} + \frac{\pi_j(1 - \pi_j)}{1 - \pi_j} \right] = \pi_k. \square$$

Example 15. With the same data as in Example 12; that is, $N = 6$, $n = 3$, and $\boldsymbol{\pi} = (0.07\ 0.17\ 0.41\ 0.61\ 0.83\ 0.91)'$, the first step of splitting for Brewer's method is given in Table 6.4.

Table 6.4. First step of Brewer's method for Example 15

	$\alpha_1 = 0.006$	$\alpha_2 = 0.015$	$\alpha_3 = 0.047$	$\alpha_4 = 0.098$	$\alpha_5 = 0.278$	$\alpha_6 = 0.555$
	$\boldsymbol{\pi}^{(1)}$	$\boldsymbol{\pi}^{(2)}$	$\boldsymbol{\pi}^{(3)}$	$\boldsymbol{\pi}^{(4)}$	$\boldsymbol{\pi}^{(5)}$	$\boldsymbol{\pi}^{(6)}$
0.07	1	0.049	0.054	0.059	0.065	0.067
0.17	0.116	1	0.131	0.142	0.157	0.163
0.41	0.280	0.290	1	0.343	0.378	0.392
0.61	0.416	0.431	0.471	1	0.562	0.584
0.83	0.567	0.587	0.641	0.695	1	0.794
0.91	0.621	0.643	0.703	0.762	0.839	1
3	3	3	3	3	3	3

At each step of the method, a unit is selected in the sample. Moreover, the $\pi_k^{(j)}$'s must not all be computed. Only the $\pi_k^{(j)}$'s of the selected vector must be computed. Practically, we can use the simple Algorithm 6.10.

The main problem of the method is that the joint inclusion probabilities are difficult to compute. Brewer (1975) proposed a recursive formula, but it implies a complete exploration of the splitting tree. The joint inclusion probabilities are, however, strictly positive because at each step, all the split vectors have strictly positive probabilities.

Algorithm 6.10 Brewer's procedure

INITIALIZE $\mathbf{s} = \mathbf{0}$.

FOR $i = 1, \ldots, n$, DO

Select a unit in the sample from U with probability

$$
p_k^{(i)} = \left[\sum_{z \in U} (1 - s_z) \frac{\pi_z \left(n - \sum_\ell s_\ell \pi_\ell - \pi_z \right)}{n - \sum_{\ell \in U} s_\ell \pi_\ell - \pi_z \{ n - (i-1) \}} \right]^{-1}
$$

$$
\times (1 - s_k) \frac{\pi_k \left(n - \sum_\ell s_\ell \pi_\ell - \pi_k \right)}{n - \sum_{\ell \in U} s_\ell \pi_\ell - \pi_k \{ n - (i-1) \}}, \quad \text{for } k \in U.
$$

IF j is the selected unit, $s_j = 1$; ENDIF.

ENDFOR.

6.3.3 Eliminatory Method

Again, we present only the first step of the method. First, define

$$
\alpha_j = \left[\sum_{z=1}^N \frac{(1 - \pi_z)(N - n - 1 + \pi_z)}{\pi_z} \right]^{-1} \frac{(1 - \pi_j)(N - n - 1 + \pi_j)}{\pi_j}.
$$

Next, compute

$$
\pi_k^{(j)} = \begin{cases} 1 - \dfrac{(1 - \pi_k)(N - n - 1)}{N - n - 1 + \pi_j} & \text{if } k \neq j \\ 0 & \text{if } k = j. \end{cases}
$$

The eliminatory method, presented in Algorithm 6.11, is the complementary design of Brewer's method. The validity follows immediately.

Algorithm 6.11 Eliminatory procedure

$\mathbf{s} = (1 \ \cdots \ 1 \ \cdots \ 1)' \in \mathbb{R}^N$.

FOR $i = 1, \ldots, N - n$, DO

Eliminate a unit from U with probability

$$
p_k^{(i)} = \left\{ \sum_{z \in U} s_z \frac{(1 - \pi_z) \left[N - n - \sum_{\ell \in U} (1 - s_\ell)(1 - \pi_\ell) - (1 - \pi_z) \right]}{N - n - \sum_{\ell \in U} (1 - s_\ell)(1 - \pi_\ell) - (1 - \pi_z) [N - n - (i-1)]} \right\}^{-1}
$$

$$
\times s_k \frac{(1 - \pi_k) \left[N - n - \sum_{\ell \in U} (1 - s_\ell)(1 - \pi_\ell) - (1 - \pi_k) \right]}{N - n - \sum_{\ell \in U} (1 - s_\ell)(1 - \pi_\ell) - (1 - \pi_k) [N - n - (i-1)]}, \quad \text{for } k \in U.
$$

IF j is the eliminated, THEN unit $s_j = 0$; ENDIF.

ENDFOR.

Example 16. With the same data as in Example 12; that is, $N = 6$, $n = 3$, and $\boldsymbol{\pi} = (0.07\ 0.17\ 0.41\ 0.61\ 0.83\ 0.91)'$, the first splitting for the eliminatory method is given in Table 6.5.

Table 6.5. First step of the eliminatory method for Example 16

π_g	$\alpha_1 = 0.624$ $\boldsymbol{\pi}^{(1)}$	$\alpha_2 = 0.240$ $\boldsymbol{\pi}^{(2)}$	$\alpha_3 = 0.079$ $\boldsymbol{\pi}^{(3)}$	$\alpha_4 = 0.038$ $\boldsymbol{\pi}^{(4)}$	$\alpha_5 = 0.013$ $\boldsymbol{\pi}^{(5)}$	$\alpha_6 = 0.007$ $\boldsymbol{\pi}^{(6)}$
0.07	0	0.143	0.228	0.287	0.343	0.361
0.17	0.198	0	0.311	0.364	0.413	0.430
0.41	0.430	0.456	0	0.548	0.583	0.595
0.61	0.623	0.641	0.676	0	0.724	0.732
0.83	0.836	0.843	0.859	0.870	0	0.883
0.91	0.913	0.917	0.925	0.931	0.936	0
3	3	3	3	3	3	3

6.3.4 Tillé's Elimination Procedure

Another elimination procedure, proposed by Tillé (1996a), is presented in Algorithm 6.12.

Algorithm 6.12 Tillé's elimination procedure

1. First, compute for sample sizes $i = n, \ldots, N$ the quantities

$$\pi(k|i) = \frac{i x_k}{\sum_{\ell \in U} x_\ell}, \tag{6.5}$$

for all $k \in U$. For any k for which (6.5) exceeds 1, set $\pi(k|i) = 1$. Next, the quantities in (6.5) are recalculated, restricted to the remaining units. This procedure is repeated until each $\pi(k|i)$ is in $[0, 1]$.
2. The steps of the algorithm are numbered in decreasing order from $N - 1$ to n. At each step, a unit k is eliminated from U with probability

$$r_{ki} = 1 - \frac{\pi(k|i)}{\pi(k|i + 1)}, \quad k \in U.$$

The inclusion probabilities can be derived directly from Algorithm 6.12. Indeed,

$$\pi_k = \prod_{i=n}^{N-1} (1 - r_{ik}),$$

and

$$\pi_{k\ell} = \prod_{i=n}^{N-1} (1 - r_{ik} - r_{i\ell}).$$

Example 17. With the elimination method, for $N = 6, n = 3$, and $\boldsymbol{\pi} = (0.07\ 0.17\ 0.41\ 0.61\ 0.83\ 0.91)'$, the matrix of joint inclusion probabilities is given by

$$
\mathbf{\Pi} = \begin{pmatrix}
0.07 & 0 & 0.0024 & 0.0265 & 0.0511 & 0.0600 \\
0 & 0.17 & 0.0058 & 0.0643 & 0.1241 & 0.1457 \\
0.0024 & 0.0058 & 0.41 & 0.1610 & 0.2994 & 0.3514 \\
0.0265 & 0.0643 & 0.1610 & 0.61 & 0.4454 & 0.5229 \\
0.0511 & 0.1241 & 0.2994 & 0.4454 & 0.83 & 0.7400 \\
0.0600 & 0.1457 & 0.3514 & 0.5229 & 0.7400 & 0.91
\end{pmatrix}. \tag{6.6}
$$

The elimination method can also be presented as a splitting technique, with the α_j's calculated as follows. First, define $F = \emptyset$. Next, repeat the following two allocations until convergence

$$
\alpha_j = \begin{cases}
1 - \pi_j \dfrac{N - 1 - \operatorname{card}(F)}{n - \sum_{i \in F} \pi_i} & \text{if } j \notin F \\
0 & \text{if } j \in F,
\end{cases}
$$
$$
F = \{j \in U | \alpha_j \le 0\}.
$$

Finally, define

$$
\pi_k^{(j)} = \begin{cases}
0 & \text{if } k = j \\
\dfrac{\pi_k}{1 - \alpha_k} & \text{if } k \ne j.
\end{cases}
$$

Note that as soon as i is small enough to get $\pi(k|i) < 0$, for all $k \in U$, the units are eliminated with equal probabilities. The problem is thus reduced to simple random sampling and the $N - n$ steps of the algorithm are rarely required. Thus, this method can be applied to large populations. Slanta and Fagan (1997), Slanta (1999), and Slanta and Kusch (2001) have modified the elimination procedure in order to avoid null joint inclusion probabilities.

Example 18. If $N = 6, n = 3$, $\boldsymbol{\pi} = (0.07 \ 0.17 \ 0.41 \ 0.61 \ 0.83 \ 0.91)'$, we obtain at the first step the splitting presented in Table 6.6. At the first step, we

Table 6.6. Elimination procedure, step 1, for Example 18

$\alpha_1 = 0.708333$	$\alpha_2 = 0.291667$
0	0.24
0.24	0
0.41	0.41
0.61	0.61
0.83	0.83
0.91	0.91

already see that π_{12} equals zero. The splitting obtained at the second step is presented in Table 6.7. At this step, three units must be selected from the four remaining units, which is straightforward.

Table 6.7. Tillé's elimination procedure, step 2 for Example 18

α_1	α_2	α_3	α_4	α_5	α_6
0.4384	0.2474	0.02249	0.1806	0.1019	0.0093
0	0	0	0	0.63	0.63
0	0.63	0.63	0	0	0
0.63	0	0.63	0.63	0	0.63
0.63	0.63	0	0.63	0.63	0
0.83	0.83	0.83	0.83	0.83	0.83
0.91	0.91	0.91	0.91	0.91	0.91

6.3.5 Generalized Midzuno Method

The Midzuno method (see Midzuno, 1950; Horvitz and Thompson, 1952; Yates and Grundy, 1953; Rao, 1963b; Avadhani and Srivastava, 1972; Chaudhuri, 1974; Korwar, 1996; Brewer and Hanif, 1983, procedure 6, p. 25) is very simple to implement and can be described as a splitting procedure into N parts that needs only one step.

The first step of the procedure is as follows. One of the vectors of $\pi_k^{(j)}$, for $j = 1, \ldots, M$, will be selected with the probabilities α_j; $\pi_k^{(j)}$ are the inclusion probabilities by means of which the selection will be applied at the next step, where

$$\pi_k^{(j)} = \begin{cases} 1 & \text{if } k = j \\ \dfrac{n-1}{N-1} & \text{if } k \neq j, \end{cases}$$

$$\alpha_j = \pi_j \frac{N-1}{N-n} - \frac{n-1}{N-n}, \quad \text{for all } j = 1, \ldots, N.$$

Because, at the second step, the problem is reduced to sampling with equal probabilities, except for one unit that is selected automatically, a simple random sampling is applied. The joint inclusion probabilities are given by

$$\pi_{k\ell} = \frac{n-1}{N-2} \left(\pi_k + \pi_\ell - \frac{n}{N-1} \right). \tag{6.7}$$

Because $\alpha_j \in [0, 1]$, the method is applicable only if

$$\pi_k \geq \frac{n-1}{N-1}, \quad \text{for all } k \in U, \tag{6.8}$$

which is very restrictive. By (6.8), we directly obtain

$$\pi_{k\ell} \geq \frac{(n-1)(n-2)}{(N-1)(N-2)} > 0.$$

The Midzuno method can be generalized to apply to any inclusion probabilities even if condition (6.8) is not satisfied. The α_j are computed by means of the following procedure. First, define $F = \emptyset$. Next, repeat the following two allocations until the same α_j, for $j = 1, \ldots, N$, are obtained in two consecutive steps:

$$\alpha_j = \begin{cases} 1 - \dfrac{(1 - \pi_j)(N - 1 - \mathrm{card}(F))}{N - n - \sum_{i \in F}(1 - \pi_i)} & \text{if } j \notin F \\ 0 & \text{if } j \in F, \end{cases}$$

$$F = \{1 \leq j \leq N | \alpha_j \leq 0\}.$$

Finally, define

$$\pi_k^{(j)} = \begin{cases} 1 & \text{if } k = j \\ \dfrac{\pi_k - \alpha_k}{1 - \alpha_k} & \text{if } k \neq j. \end{cases}$$

Moreover, when only one iteration is needed, the algorithm provides the α_j of Midzuno's method. The fundamental difference from the classical Midzuno method is that the problem is not necessarily reduced to a simple random sampling at the second step. The algorithm is thus repeated until an equal probability vector is obtained, which allows applying a simple random sampling.

Example 19. With the same data as in Example 12; that is, $N = 6, n = 3$, and $\boldsymbol{\pi} = (0.07\ 0.17\ 0.41\ 0.61\ 0.83\ 0.91)$, at the first step, $\alpha_i = 0, i = 1, \ldots, 4$, $\alpha_5 = 0.346$, and $\alpha_6 = 0.654$. The problem is thus reduced to a splitting into two parts; see Table 6.8, step 1.

Table 6.8. Generalized Midzuno procedure for Example 19

Step 1		Step 2					
$\alpha_5 = 0.346$	$\alpha_6 = 0.654$	0.017	0.128	0.201	0.032	0.243	0.380
0.07	0.07	0.07	0.07	0.07	0.07	0.07	0.07
0.17	0.17	0.17	0.17	0.17	0.17	0.17	0.17
0.41	0.41	1	0.38	0.38	1	0.38	0.38
0.61	0.61	0.38	1	0.38	0.38	1	0.38
1	0.74	1	1	1	0.38	0.38	1
0.74	1	0.38	0.38	1	1	1	1

At the second step, the method provides a splitting of each of these two parts into three parts. Finally, we get the breakdown given in Table 6.8, step 2. At step 3, the problem consists of selecting only one unit. This example shows, however, that the generalized Midzuno method does not ensure strictly positive joint inclusion probabilities, because $\pi_{12} = 0$.

Let $p^c(\mathbf{s})$ denote the complementary design (see Definition 16, page 14) of a sampling design $p(\mathbf{s})$. If the design $p(\mathbf{s})$ has the inclusion probabilities π_k, then $p^c(U \backslash s) = p(\mathbf{s})$, $\pi_k^c = 1 - \pi_k$, $k \in U$, and

$$\pi_{k\ell}^c = 1 - \pi_k - \pi_\ell + \pi_{k\ell}, k \neq \ell \in U.$$

It is easy to see that the generalized Midzuno method is complementary to Tillé's elimination method. The simple draw by draw Algorithm 6.13 can thus be derived from Tillé's elimination procedure.

Algorithm 6.13 Generalized Midzuno procedure
___

1. Compute $\pi_k^c = 1 - \pi_k, k \in U$.
2. First, compute for sample sizes $i = N - n, \ldots, N$ the quantities

$$\pi_k^c(k|i) = \frac{i\pi_k^c}{\sum_{\ell \in U} x_\ell}, \qquad (6.9)$$

   for all $k \in U$. For any k for which (6.9) exceeds 1, set $\pi^c(k|i) = 1$. Next, the quantities in (6.9) are recalculated, restricted to the remaining units. This procedure is repeated until each $\pi^c(k|i)$ is in $[0,1]$.
3. The steps of the algorithm are numbered in decreasing order from $N - 1$ to $N - n$. At each step, a unit k is selected from U with probability

$$p_{ki} = 1 - \frac{\pi(k|i)}{\pi(k|i+1)}, \quad k \in U.$$

6.3.6 Chao's Method

For the original description of this reservoir procedure, see Chao (1982) and Richardson (1989), but Chao's method was also studied by Sengupta (1989), Bethlehem and Schuerhoff (1984), Sugden et al. (1996), and Berger (1998b). The Chao method presented in Algorithm 6.14 is a generalization of the reservoir method for SRSWOR (see Algorithm 4.4, page 49).

It is also possible to present an arguably simpler description of the method as an elimination procedure. At each of the $N - n$ steps of the algorithm, a unit is eliminated from the first $n + 1$ units of the sample. The method can thus be presented as a splitting technique into $M = n + 1$ parts. First, define $F = \emptyset$. Next, repeat the following two allocations until convergence:

$$\alpha_j = \begin{cases} 1 - \pi_j \dfrac{n - \operatorname{card}(F)}{\sum_{i=1 \setminus F}^{n+1} \pi_i} & \text{if } j \notin F \\ 0 & \text{if } j \in F, \end{cases}$$

$$F = \{1 \leq j \leq n + 1 | \alpha_j \leq 0\}.$$

Then, define

$$\pi_k^{(j)} = \begin{cases} 0 & \text{if } k = j \\ \dfrac{\pi_k}{1 - \alpha_k} & \text{if } k \neq j, k = 1, \ldots, n + 1, \\ \pi_k & \text{if } k = n + 2, \ldots, N. \end{cases}$$

Algorithm 6.14 Chao's procedure

1. First, compute for sample sizes $i = n, \dots, N$ the quantities

$$\pi(k|i) = \frac{n x_k}{\sum_{\ell=1}^{i} x_\ell}, \tag{6.10}$$

   for all $k = 1, \dots, i$. For any k for which (6.10) exceeds 1, we set $\pi(k|i) = 1$. Next, the quantities in (6.10) are recalculated, restricted to the remaining units. This procedure is repeated until each $\pi(k|i)$ is in $[0, 1]$.

2. Select the first n units of the population in the initial sample.

3. FOR $j = n + 1, \dots, N$ DO

   Generate a random variable with a uniform distribution $u \sim \mathcal{U}[0, 1]$;
   IF $u < \pi(j|j)$ THEN
     select unit j;
     remove a unit from the sample with probability

$$\frac{1}{\pi(j|j)} \left[1 - \frac{\pi(k|j)}{\pi(k|j-1)} \right],$$

   for all the units k that belong to the sample and replace it by unit j;
   ELSE retain the old sample;
   ENDIF;
   ENDFOR.

6.4 Splitting Procedures in Brewer and Hanif

The splitting method allows constructing an infinite number of methods. In Brewer and Hanif (1983), at least five splitting procedures can be written directly under the form of a splitting procedure. The three methods proposed by Jessen (1969) are typically particular cases of the splitting method (see Algorithm 6.4, p. 103), they are linked to the minimum support design. The Brewer procedure (see also Brewer, 1963a, 1975; Brewer and Hanif, 1983, procedure 8, page 26) is maybe best presented in the form of a splitting method (see Algorithm 6.10, page 114). These methods are summarized in Table 6.9, page 121.

Table 6.9. Brewer and Hanif list of splitting procedures

Number	Page	Procedure name	strpps	strwor	n fixed	syst	d by d	rej ws	ord	unord	inexact	n=2 only	b est var	j p enum	j p iter	not gen app	nonrotg	exact	exponential
		Midzuno's Procedure																	
6	25	Principal reference: Horvitz and Thompson (1952)	+	+	+	–	+	–	–	–	–	–	–	–	+	+	+	–	–
		Other ref.: Yates and Grundy (1953), Rao (1963b),																	
		Brewer's procedure																	
8	26	Principal reference: Brewer (1975), Brewer (1963a), Brewer (1963b)	+	+	+	–	+	–	–	–	–	–	+	–	–	+	+	+	–
		Other ref.: Rao and Bayless (1969), Rao and Singh (1973)																	
		Sadasivan and Sharma (1974), Cassel et al. (1993, p. 16)																	
		Carroll-Hartley Draw by Draw Procedure																	
15	31	Principal reference: Carroll and Hartley (1964)	+	+	+	–	+	–	–	–	–	–	–	–	+	+	+	–	+
		Other ref.: Cassel et al. (1993, p. 16)																	
34	41	Jessen's "Method 1" Principal reference: Jessen (1969)	+	+	+	–	+	–	–	–	–	–	–	–	+	+	+	–	–
35	42	Jessen's "Method 2" Principal reference: Jessen (1969)	+	+	+	–	–	–	–	–	–	–	–	–	+	+	+	–	–
36	43	Jessen's "Method 3" Principal reference: Jessen (1969)	+	+	+	–	+	–	–	–	–	–	–	–	+	+	+	–	–
37	45	Jessen's "Method 4" Principal reference: Jessen (1969)	+	+	+	–	–	+	–	–	–	+	–	–	–	+	+	–	–
40	46	Das-Mohanty Procedure Principal reference: Das and Mohanty (1973)	+	+	+	–	–	+	–	–	+	+	–	–	–	–	–	–	–

7

More on Unequal Probability Designs

As we have seen in Chapters 5 and 6, there exist a large number of sampling methods with unequal probabilities. Most of them can be expressed in the form of a splitting method. Brewer and Hanif (1983) listed 50 methods in chronological order. However, only 20 of them are really "exact" in the sense that they are without replacement, have a fixed sample size, can be applied to any vector of inclusion probabilities, and respect the fixed inclusion probabilities. The exact methods are listed in Table 7.1, page 125. Several methods of this list have already been described in Chapters 5 and 6.

In this chapter, we present some methods that are neither exponential nor a particular case of the splitting procedure. Four methods have a real practical interest and are presented in this chapter: ordered systematic sampling, random systematic sampling, Deville's systematic sampling and Sampford's method. Each one can be easily implemented.

Next, we discuss a way to approximate the variance of the Horvitz-Thompson estimator in an exponential design or in a design close to an exponential design. We have seen that several proposed algorithms are such that it is impossible to compute the corresponding sampling design. It is not necessary to specify the sampling design for estimating simple population statistics such as totals, means, and ratios. For that purpose, knowledge of the first-order inclusion probabilities is all that is required. In Sections 7.5, page 137 and 8.8, page 169, we show that it is even possible to estimate the variances of the estimates given no more than those first-order inclusion probabilities. So, being able to specify the sample design merely in order to evaluate the second-order inclusion probabilities, once considered essential for the estimation of variance, is now of little or no consequence.

We propose several approximations of variance that allow constructing several variance estimators. Although biased, these estimators generally have a smaller mean square error than the Horvitz-Thompson estimators or the Sen-Yates-Grundy estimators of variance. Finally, the variance-covariance operators of the four sampling designs are compared to two approximations of variance, which allow choosing a method of sampling with unequal probabilities and an approximation of the variance.

7.1 Ordered Systematic Sampling

Systematic sampling was first proposed by Madow (1949) and is one of the most useful methods owing to its simplicity and its exactness. Suppose that the inclusion probabilities are such that $0 < \pi_k < 1, k \in U$ with

$$\sum_{k \in U} \pi_k = n.$$

Define

$$V_k = \sum_{\ell=1}^{k} \pi_\ell, \text{ for all } k \in U, \tag{7.1}$$

with $V_0 = 0$ and $V_N = n$.

The method is the following. First, generate u, a uniform variable $\mathcal{U}[0, 1]$.

- The first selected unit k_1 is such that $V_{k_1-1} \le u < V_{k_1}$,
- The second selected unit k_2 is such that $V_{k_2-1} \le u + 1 < V_{k_2}$,
- The jth selected unit is such that $V_{k_j-1} \le u + j - 1 < V_{k_j}$.

More formally, systematic sampling can be defined as follows. Select the units $k_i, i = 1, \ldots, n$, such that the intervals $[V_{k-1} - u, V_k - u[$ contain an integer number. This definition allows constructing Algorithm 7.1 that is particularly simple, where the notation $\lfloor x \rfloor$ denotes the largest integer number smaller than x.

Algorithm 7.1 Systematic sampling

DEFINITION a, b, u real; k INTEGER;
$u = \mathcal{U}[0, 1[$;
$a = -u$;

FOR $k = 1, \ldots, N$ DO
  $b = a$;
  $a = a + \pi_k$;
  IF $\lfloor a \rfloor \ne \lfloor b \rfloor$ THEN select k ENDIF;
ENDFOR.

However, ordered systematic sampling has a drawback: a lot of joint inclusion probabilities can be equal to zero. Formally, it is possible to derive a general expression of the joint inclusion probabilities that depends on the quantities:

Table 7.1. Brewer and Hanif (1983) list of the exact sampling procedures. The abbreviations are given in Table 5.6, page 96

Number	Page	Procedure name	strpps	strwor	n fixed	syst	d by d	rej ws	ord	unord	inexact	n=2 only	b est var	j p enum	j p iter	not gen app	nonrotg	exact	exponential
1	21	Ordered Systematic Procedure, Madow (1949)	+	+	+	+	+	−	+	−	−	−	+	+	−	−	−	+	−
2	22	Random Systematic Procedure, Goodman and Kish (1950)	+	+	+	+	+	−	−	−	−	−	+	+	−	−	−	+	−
3	23	Grundy's Systematic Procedure, Grundy (1954)	+	+	+	+	+	−	−	−	−	−	+	+	−	−	−	+	−
8	26	Brewer's Procedure, Brewer (1975, 1963a,b)	+	+	+	+	−	+	−	+	−	−	+	+	−	−	+	+	−
11	28	Rao-Sampford Rejective Procedure, Rao (1965), Sampford (1967)	+	+	+	+	−	+	−	+	−	−	−	−	+	−	+	+	−
12	29	Durbin-Sampford Procedure, Sampford (1967)	+	+	+	+	−	+	−	+	−	−	−	−	+	−	+	+	−
13	30	Fellegi's Procedure, Fellegi (1963)	+	+	+	+	−	+	−	+	−	−	−	−	+	−	−	+	−
14	31	Carroll-Hartley Rejective Procedure, Carroll and Hartley (1964)	+	+	+	+	−	+	−	+	−	−	−	−	+	−	+	+	+
15	31	Carroll-Hartley Draw by Draw Procedure, Carroll and Hartley (1964)	+	+	+	+	+	−	−	+	−	−	−	−	+	−	+	+	−
16	32	Carroll-Hartley Whole Sample Procedure, Carroll and Hartley (1964)	+	+	+	+	+	−	−	+	−	−	−	−	+	−	+	+	−
18	33	Hanurav's Scheme B-A, Hanurav (1967)	+	+	+	+	+	−	−	+	−	−	−	−	+	−	+	+	−
19	34	Hanurav-Vijayan Procedure, Hanurav (1967), Vijayan (1968)	+	+	+	+	+	−	−	+	−	−	−	−	−	+	+	+	−
34	41	Jessen's "Method 1", Jessen (1969)	+	+	+	+	−	−	−	+	−	+	−	−	−	−	+	+	−
35	42	Jessen's "Method 2", Jessen (1969)	+	+	+	+	−	−	−	+	−	+	−	−	−	−	+	+	−
36	43	Jessen's "Method 3", Jessen (1969)	+	+	+	+	−	−	−	+	−	+	−	−	−	−	+	+	−
41	46	Mukhopadhyay's Procedure, Mukhopadhyay (1972)	+	+	+	+	−	−	−	+	−	−	−	−	−	−	+	+	−
42	47	Sinha's Extension Procedure, Sinha (1973)	+	+	+	+	−	−	−	+	−	−	−	−	−	−	+	+	−
43	47	Sinha's Reduction Procedure, Sinha (1973)	+	+	+	+	−	−	−	+	−	−	−	−	−	−	+	+	−
49	50	Choudhry's Procedure, Choudhry (1979)	+	+	+	+	−	−	+	−	−	−	−	−	−	−	+	+	−
50	51	Chromy's Procedure, Chromy (1979)	+	+	+	+	+	−	+	−	−	−	−	−	+	−	+	+	−

$$V_{k\ell} = \begin{cases} \displaystyle\sum_{i=k}^{\ell-1} \pi_i, & \text{if } k < \ell \\[2ex] \displaystyle\sum_{i=k}^{N} \pi_i + \sum_{i=1}^{\ell-1} \pi_i = n - \sum_{i=\ell}^{k-1} \pi_i, & \text{if } k > \ell. \end{cases}$$

The joint inclusion probabilities are given by:

$$\pi_{k\ell} = \min\left\{\max\left(0, \pi_k - \delta_{k\ell}\right), \pi_\ell\right\} + \min\left\{\pi_k, \max\left(0, \delta_{k\ell} + \pi_\ell - 1\right)\right\}, k < \ell, \tag{7.2}$$

where $\delta_{k\ell} = V_{k\ell} - \lfloor V_{k\ell} \rfloor$ (for further results, see Connor, 1966; Pinciaro, 1978; Hidiroglou and Gray, 1980).

Example 20. Suppose that $N = 6$ and $n = 3$. The inclusion probabilities and the cumulative inclusion probabilities are given in Table 7.2. Suppose also

Table 7.2. Inclusion probabilities and cumulative probabilities for Example 20

k	0	1	2	3	4	5	6	Total
π_k	0	0.07	0.17	0.41	0.61	0.83	0.91	3
V_k	0	0.07	0.24	0.65	1.26	2.09	3	

that the value taken by the uniform random number is $u = 0.354$. The rules of selections presented in Figure 7.1 are:

- Because $V_2 \le u < V_3$, unit 3 is selected;
- Because $V_4 \le u < V_5$, unit 5 is selected;
- Because $V_5 \le u < V_6$, unit 6 is selected.

The sample selected is thus $\mathbf{s} = (0, 0, 1, 0, 1, 1)$.

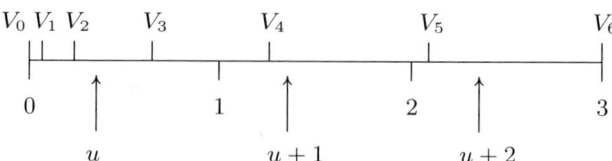

Fig. 7.1. Rules of selection in systematic sampling

An examination of Figure 7.1 allows deriving completely the sampling design:

- If $0 \le u < 0.07$ then the selected sample is $(1, 0, 0, 1, 1, 0)$;

- If $0.07 \leq u < 0.09$ then the selected sample is $(0, 1, 0, 1, 1, 0)$;
- If $0.09 \leq u < 0.24$ then the selected sample is $(0, 1, 0, 1, 0, 1)$;
- If $0.24 \leq u < 0.26$ then the selected sample is $(0, 0, 1, 1, 0, 1)$;
- If $0.26 \leq u < 0.65$ then the selected sample is $(0, 0, 1, 0, 1, 1)$;
- If $0.65 \leq u < 1$ then the selected sample is $(0, 0, 0, 1, 1, 1)$.

This result allows deriving the sampling design in Table 7.3.

Table 7.3. Sampling design obtained by systematic sampling in Example 20

k	s_1	s_2	s_3	s_4	s_4	s_6	π
1	1	0	0	0	0	0	0.07
2	0	1	1	0	0	0	0.17
3	0	0	0	1	1	0	0.41
4	1	1	1	1	0	1	0.61
5	1	1	0	0	1	1	0.83
6	0	0	1	1	1	1	0.91
$p(\mathbf{s})$	0.07	0.02	0.15	0.02	0.39	0.35	

Finally, the joint inclusion probabilities can be computed, either from Table 7.3 or from general Expression (7.2).

$$
\Pi = \begin{pmatrix}
0.07 & 0 & 0 & 0.07 & 0.07 & 0 \\
0 & 0.17 & 0 & 0.17 & 0.02 & 0.15 \\
0 & 0 & 0.41 & 0.02 & 0.39 & 0.41 \\
0.07 & 0.17 & 0.02 & 0.61 & 0.44 & 0.52 \\
0.07 & 0.02 & 0.39 & 0.44 & 0.83 & 0.74 \\
0 & 0.15 & 0.41 & 0.52 & 0.74 & 0.91
\end{pmatrix}.
$$

Several inclusion probabilities are null. Under certain assumptions, it is, however, possible to construct a variance estimator under ordered systematic sampling (see among others Madow and Madow, 1944; Cochran, 1946; Bellhouse and Rao, 1975; Bellhouse, 1988; Iachan, 1982, 1983; Berger, 2003). Péa and Tillé (2005) have shown that systematic sampling is a minimum support design.

7.2 Random Systematic Sampling

In order to overcome the problem of null joint inclusion probabilities, systematic sampling can be applied after randomly sorting the file. With ordered systematic sampling, the joint inclusion probabilities depend on the order of the units. So the joint inclusion probabilities of random systematic sampling are the means of the joint inclusion probabilities of ordered systematic sampling for all permutations of the population. The computation is thus confronted with the problem of combinatorial explosion but can be carried out for a small population size.

Example 21. With the random systematic method, for $N = 6, n = 3$, and $\boldsymbol{\pi} = (0.07\ 0.17\ 0.41\ 0.61\ 0.83\ 0.91)'$, the matrix of joint inclusion probabilities is given by

$$
\boldsymbol{\Pi} = \begin{pmatrix}
0.07 & 0.0140 & 0.0257 & 0.0257 & 0.0373 & 0.0373 \\
0.0140 & 0.17 & 0.0623 & 0.0623 & 0.0740 & 0.1273 \\
0.0257 & 0.0623 & 0.41 & 0.0873 & 0.2957 & 0.3490 \\
0.0257 & 0.0623 & 0.0873 & 0.61 & 0.4957 & 0.5490 \\
0.0373 & 0.0740 & 0.2957 & 0.4957 & 0.83 & 0.7573 \\
0.0373 & 0.1273 & 0.3490 & 0.5490 & 0.7573 & 0.91
\end{pmatrix}. \tag{7.3}
$$

Nevertheless, a random sorting of the data does not entirely solve the problem of null joint inclusion probabilities. Consider the following example.

Example 22. Suppose that $N = 5$, $n = 2$, and $\pi_1 = \pi_2 = 0.25, \pi_3 = \pi_4 = \pi_5 = 0.5$. With systematic sampling, the probability of jointly selecting the "smallest" two units is null whatever the order of the units. Indeed, for random systematic sampling, the matrix of joint inclusion probabilities is

$$
\boldsymbol{\Pi} = \begin{pmatrix}
0.250 & 0 & 0.083 & 0.083 & 0.083 \\
0 & 0.250 & 0.083 & 0.083 & 0.083 \\
0.083 & 0.083 & 0.500 & 0.170 & 0.170 \\
0.083 & 0.083 & 0.170 & 0.500 & 0.170 \\
0.083 & 0.083 & 0.170 & 0.170 & 0.500
\end{pmatrix}.
$$

Hartley and Rao (1962) proposed an approximation of the joint inclusion probabilities based on an Edgeworth expansion of the distribution function of the $V_{k\ell}$. However, a large number of publications have shown that the computation of the joint inclusion probabilities are not necessary to estimate the variance (see among others Hartley and Rao, 1962; Raj, 1965; Wolter, 1984; Rosén, 1991; Stehman and Overton, 1994; Brewer and Donadio, 2003; Matei and Tillé, 2006). The general issue of variance estimation in unequal probability sampling is developed in Section 7.5.2, page 140.

7.3 Deville's Systematic Sampling

Deville (1998) presented a particular systematic technique. The V_k's are constructed according to Expression (7.1). Deville's technique consists of selecting only one unit at random in each interval $[i - 1, i[$ with $i = 1, \ldots, n$. The technique consists of generating n random variables $u_i, i = 1, \ldots, n$, with a uniform distribution $u_i \sim \mathcal{U}[0, 1[$. For each random variable i, unit k is selected if

$$V_{k-1} \leq u_i + (i-1) < V_k.$$

Suppose that we want to select four units in a population of size $N = 7$. Figure 7.2 illustrates Deville's method.

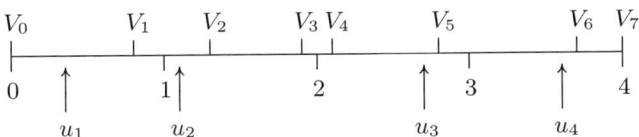

Fig. 7.2. Deville's method

Figure 7.2 shows that after generating u_1, u_2, u_3, and u_4, units 1, 2, 5, and 6 are selected. An examination of the example shows that an intricate problem must be considered. A unit ℓ is said to be cross-border if $[V_{\ell-1}, V_\ell[$ contains an integer number. In Figure 7.2, units 2, 4, and 6 are cross-border. If the u_i are selected independently, the cross-border units can be selected twice. Deville proposed to introduce a dependence between the u_i in such a way that no unit is selected twice.

First, let ℓ be the cross-border unit such that $[V_{\ell-1} \leq i - 1 < V_\ell[$. In this case, unit ℓ can be selected at step $i - 1$. The density function of u_i is defined in such a way that ℓ is not selected twice:

- If ℓ is selected at step $i - 1$, u_i has the density function:

$$f_1(x) = \begin{cases} \dfrac{1}{i - V_\ell} & \text{if } x \geq V_\ell - (i-1) \\ 0 & \text{if } x < V_\ell - (i-1) \end{cases}, x \in [0,1[.$$

- If ℓ is not selected at step $i - 1$, u_i has the density function:

$$f_2(x) = \begin{cases} 1 - \dfrac{(i - 1 - V_{\ell-1})(V_\ell - i + 1)}{[1 - (i - 1 - V_{\ell-1})][1 - (V_\ell - i + 1)]} & \text{if } x \geq V_\ell - i + 1 \\ \dfrac{1}{1 - (i - 1 - V_{\ell-1})} & \text{if } x < V_\ell - i + 1. \end{cases}$$

Knowing that the probability of selecting unit ℓ at step $i - 1$ is equal to $i - 1 - V_{\ell-1}$, it can be shown that u_i has a uniform distribution in $[0,1[$. Indeed,

$$(i - 1 - V_{\ell-1})f_1(x) + [1 - (i - 1 - V_{\ell-1})]f_2(x) = 1, x \in [0,1[.$$

Deville's method is presented in Algorithm 7.2. The computation of the joint inclusion probabilities is tedious and is developed in Deville (1998).

Algorithm 7.2 Deville's systematic procedure

---

1. Generate u_1, a realization of a uniform random variable in [0,1]. Unit k such that $V_{k-1} \le u_1 < V_k$ is selected.
2. FOR $i = 2, \ldots, n$, DO

   Let ℓ be the cross-border unit such that $[V_{\ell-1} \le i - 1 < V_\ell[$.

   IF unit ℓ is selected at step $i - 1$ THEN

$$f(x) = \begin{cases} \dfrac{1}{i - V_\ell} & \text{if } x \ge V_\ell - (i - 1) \\ 0 & \text{if } x < V_\ell - (i - 1) \end{cases}, x \in [0, 1[.$$

   ELSE

$$f(x) = \begin{cases} 1 - \dfrac{(i - 1 - V_{\ell-1})(V_\ell - i + 1)}{[1 - (i - 1 - V_{\ell-1})][1 - (V_\ell - i + 1)]} & \text{if } x \ge V_\ell - i + 1 \\ \dfrac{1}{1 - (i - 1 - V_{\ell-1})} & \text{if } x < V_\ell - i + 1. \end{cases}$$

   ENDIF

   Generate u_i, a random variable with density $f(x)$.

   Unit k such that $V_{k-1} \le u_i + i - 1 < V_k$ is selected.

   ENDFOR

7.4 Sampford Rejective Procedure

7.4.1 The Sampford Sampling Design

Sampford's method (1967) is a very ingenious procedure (see also Bayless and Rao, 1970; Asok and Sukhatme, 1976; Gabler, 1981). It is linked with the exponential methods but is not formally exponential. The implementation is very simple, and the joint inclusion probabilities can be computed relatively easily.

Let $\pi_1, \ldots, \pi_k, \ldots, \pi_N$ be given inclusion probabilities such that

$$\sum_{k \in U} \pi_k = n$$

and

$$\omega_k = \frac{\pi_k}{1 - \pi_k}.$$

The Sampford sampling design is defined as follows:

$$p_{\text{SAMPFORD}}(\mathbf{s}) = C_n \sum_{\ell=1}^{N} \pi_\ell s_\ell \prod_{\substack{k=1 \\ k \neq \ell}}^{N} \omega_k^{s_k}, \tag{7.4}$$

or equivalently

$$p_{\text{SAMPFORD}}(\mathbf{s}) = C_n \left(\prod_{k=1}^{N} \omega_k^{s_k} \right) \left(n - \sum_{\ell=1}^{N} \pi_\ell s_\ell \right),$$

where

$$C_n = \left(\sum_{t=1}^{n} t D_{n-t} \right)^{-1}$$

and

$$D_z = \sum_{\mathbf{s} \in \mathcal{S}_z} \prod_{k=1}^{N} \omega_k^{s_k},$$

for $z = 1, \ldots, N$, and $D_0 = 1$.

7.4.2 Fundamental Results

In order to prove that

$$\sum_{\mathbf{s} \in \mathcal{S}_n} p_{\text{SAMPFORD}}(\mathbf{s}) = 1$$

and that the sampling design reaches the required inclusion probabilities, Sampford first proved the following lemma:

Lemma 1. *(Sampford, 1967) If*

$$g(n, r, k) = \sum_{\mathbf{s} \in \mathcal{S}_{n-r}(U \backslash k)} \left(\prod_{\ell \in U} \omega_\ell^{s_\ell} \right) \left(n - r\pi_k - \sum_{i=1}^{N} \pi_i s_i \right),$$

where

$$\mathcal{S}_{n-r}(U \backslash \{k\}) = \{\mathbf{s} \in \mathcal{S}_{n-r} | s_k = 0\},$$

then

$$g(n, r, k) = (1 - \pi_k) \sum_{t=r}^{n} t D_{n-t}. \tag{7.5}$$

Proof. When $\text{card}(\mathbf{s}) = n - r$,

$$n - r\pi_k - \sum_{i=1}^{N} \pi_i s_i = r(1 - \pi_k) + \sum_{i=1}^{N} (1 - \pi_i) s_i.$$

Thus

$$g(n, r, k) = \left[r(1 - \pi_k) D_{n-r}(\bar{k}) + h(n, r, k) \right], \tag{7.6}$$

where

$$h(n, r, k) = \sum_{s \in \mathcal{S}_{n-r}(U \setminus k)} \left(\prod_{\ell \in U} \omega_\ell^{s_\ell} \right) \sum_{i=1}^{N} (1 - \pi_i) s_i, \tag{7.7}$$

and

$$D_z(\bar{k}) = \sum_{s \in \mathcal{S}_z(U \setminus \{k\})} \prod_{\ell=1}^{N} \omega_\ell^{s_\ell}.$$

Note that $1 - \pi_i = \pi_i/\omega_i$. Now, if in Expression (7.7), we replace $1 - \pi_i$ with π_i/ω_i, we get

$$h(n, r, k) = \sum_{s \in \mathcal{S}_{n-r}(U \setminus k)} \left(\prod_{\ell \in U} \omega_\ell^{s_\ell} \right) \sum_{i=1}^{N} \frac{\pi_i}{\omega_i} s_i$$

$$= \sum_{s \in \mathcal{S}_{n-r-1}(U \setminus k)} \left(\prod_{\ell \in U} \omega_\ell^{s_\ell} \right) \left[\sum_{\substack{i=1 \\ i \neq k}}^{N} \pi_i (1 - s_i) \right]$$

$$= \sum_{s \in \mathcal{S}_{n-r-1}(U \setminus k)} \left(\prod_{\ell \in U} \omega_\ell^{s_\ell} \right) \left(n - \pi_k - \sum_{\substack{i=1 \\ i \neq k}}^{N} \pi_i s_i \right)$$

$$= g(n, r+1, k) + r \pi_k D_{n-r-1}(\bar{k}). \tag{7.8}$$

Because

$$D_m(\bar{k}) = D_m - \omega_k D_{m-1}(\bar{k}),$$

by Expressions (7.6) and (7.8), we obtain the recursive relation:

$$g(n, r, k) = r(1 - \pi_k) D_{n-r} + g(n, r+1, k),$$

with the initial condition

$$g(n, n, k) = 1 - \pi_k.$$

Expression (7.5) satisfies this recurrence relation. □

Lemma 1 also holds when $\pi_k > 1$. In the case where $\pi_k = 1$, Sampford (1967) pointed out that a modified form can be stated:

$$g(n, r, k) = \pi_k \sum_{t=r}^{n-1} D_{n-t-1}(\bar{k}). \tag{7.9}$$

Now, the most fundamental results can be derived from Lemma 1.

Result 31. *(Sampford, 1967) For the Sampford design defined in (7.4)*

(i) $\displaystyle\sum_{s \in \mathcal{S}_n} p_{\text{SAMPFORD}}(s) = 1.$

(ii) $\displaystyle\sum_{s \in \mathcal{S}_n} s_k p_{\text{SAMPFORD}}(s) = \pi_k.$

(iii) $\displaystyle \pi_{k\ell} = C_n \omega_k \omega_\ell \sum_{t=2}^{n} (t - \pi_k + \pi_\ell) D_{n-t}(\bar{k}, \bar{\ell}),$

where

$$D_z(\bar{k}, \bar{\ell}) = \sum_{s \in \mathcal{S}_z(U \backslash \{k, \ell\})} \prod_{j=1}^{N} \omega_j^{s_j}.$$

Proof.
(i) Suppose that a phantom unit z has a null inclusion probability; that is, $\pi_z = \omega_z = 0$. Then, by Lemma 1:

$$\sum_{s \in \mathcal{S}_n} p_{\text{SAMPFORD}}(s) = C_n g(n, 0, z) = C_n \sum_{t=1}^{n} t D_{n-t} = 1.$$

(ii) $\displaystyle\sum_{s \in \mathcal{S}_n} s_k p_{\text{SAMPFORD}}(s) = C_n \omega_k g(n, 1, k) = \pi_k C_n \sum_{t=1}^{n} t D_{n-t} = \pi_k.$

(iii) $\displaystyle\sum_{s \in \mathcal{S}_n} s_k s_\ell p_{\text{SAMPFORD}}(s) = C_n \omega_k \omega_\ell \Psi_{k\ell},$ $\hspace{2cm}$ (7.10)

where

$$\Psi_{k\ell} = \sum_{s \in \mathcal{S}_{n-2}(U \backslash \{k, \ell\})} \prod_{i=1}^{N} \omega_i^{s_i} \left(n - \pi_k - \pi_\ell - \sum_{j=1}^{N} \pi_j s_j \right).$$

Now suppose that units k and ℓ are aggregated in a new unit α; that is, the size of this new population U' is $N - 1$, $\pi_\alpha = \pi_k + \pi_\ell$, and $\omega_\alpha = \pi_\alpha/(1 - \pi_\alpha)$. Moreover, define g' and D' as the functions g and L defined over the modified population U'. Then

$$\Psi_{k\ell} = \sum_{s \in \mathcal{S}_{n-2}(U' \backslash \{\alpha\})} \prod_{i=1}^{N} \omega_i^{s_i} \left(n - \pi_\alpha - \sum_{j=1}^{N} \pi_j s_j \right)$$
$$= g'(n, 2, \alpha) + \pi_\alpha D'_{n-2}(\bar{\alpha}).$$

Because

$$D'_{n-t} = D'_{n-t}(\bar{\alpha}) + \omega_\alpha D'_{n-t-1}(\bar{\alpha}),$$

for $t < n - 1$ with $D'_0(\bar{\alpha}) = 1$, using Lemma 1, we obtain

$$\Psi_{k\ell} = (1 - \pi_\alpha) \left[\sum_{t=2}^{n} t D'_{n-t}(\bar{\alpha}) + \sum_{t=2}^{n} t \omega_\alpha D'_{n-t-1}(\bar{\alpha}) \right] + \pi_\alpha D'_{n-2}(\bar{\alpha})$$

$$= \sum_{t=2}^{n} (t - \pi_\alpha) D'_{n-t}(\bar{\alpha}).$$

The case where $\pi_\alpha = 1$ can be treated by using Expression (7.9). Reverting to the original population, we obtain

$$\Psi_{k\ell} = \sum_{t=2}^{n} (t - \pi_k + \pi_\ell) D_{n-t}(\bar{k}, \bar{\ell}). \tag{7.11}$$

The proof is completed by inserting (7.11) in (7.10).   □

7.4.3 Technicalities for the Computation of the Joint Inclusion Probabilities

In order to compute the joint inclusion probabilities, it is not necessary to enumerate all the possible samples. Sampford pointed out that the following relation, which is proved in Result 25, page 88, can be used,

$$D_m = \frac{1}{m} \sum_{r=1}^{m} (-1)^{r-1} \left(\sum_{i=1}^{N} \omega^r \right) D_{m-r},$$

with the initial condition $D_0 = 1$. Next, a recursive relation can be used to compute $D_m(\bar{k}, \bar{\ell})$:

$$D_m(\bar{k}, \bar{\ell}) = D_m - (\omega_k + \omega_\ell) D_{m-1}(\bar{k}, \bar{\ell}) - \omega_k \omega_\ell D_{m-2}(\bar{k}, \bar{\ell}).$$

Gabler (1981) has shown that Sampford's procedure has a smaller variance than multinomial sampling; that is, sampling with unequal probabilities with replacement.

Example 23. With Sampford's method, for $N = 6, n = 3$, and

$$\pi = (0.07\ 0.17\ 0.41\ 0.61\ 0.83\ 0.91)',$$

the matrix of joint inclusion probabilities is given by

$$\Pi = \begin{pmatrix} 0.07 & 0.004 & 0.012 & 0.022 & 0.046 & 0.057 \\ 0.004 & 0.17 & 0.029 & 0.054 & 0.114 & 0.139 \\ 0.012 & 0.029 & 0.41 & 0.145 & 0.289 & 0.345 \\ 0.022 & 0.054 & 0.145 & 0.61 & 0.466 & 0.533 \\ 0.046 & 0.114 & 0.289 & 0.466 & 0.83 & 0.745 \\ 0.057 & 0.139 & 0.345 & 0.533 & 0.745 & 0.91 \end{pmatrix}. \tag{7.12}$$

7.4.4 Implementation of the Sampford Design

General result

The most powerful algorithms for implementing Sampford's design are based on the following result.

Result 32. *Let $\mathcal{Q}$ be a support such that $\mathcal{S}_{n-1} \subset \mathcal{Q}$. Let $\mathbf{S}_1$ and $\mathbf{S}_2$ be two independent random samples such that*

- $\mathbf{S}_1 = (S_{11} \cdots S_{1N})'$ *is a random sample of size 1 and inclusion probabilities $q_k = \pi_k/n$ from U,*
- $\mathbf{S}_2 = (S_{21} \cdots S_{2N})'$ *is an exponential random sample with support $\mathcal{Q}$ with parameter $\boldsymbol{\lambda} = (\lambda_1 \cdots \lambda_N)'$ and $\lambda_k = \log(\omega_k) = \log[\pi_k/(1-\pi_k)]$.*

Then $\Pr[\mathbf{S}_1 + \mathbf{S}_2 = \mathbf{s} | (\mathbf{S}_1 + \mathbf{S}_2) \in \mathcal{S}_n]$ is a Sampford sampling design with inclusion probabilities $\pi_k, k \in U$.

Proof. Because

$$\Pr[\mathbf{S}_1 = \mathbf{s}] = \prod_{k \in U} q_k^{s_k}, \text{ for all } \mathbf{s} \in \mathcal{S}_1,$$

and

$$\Pr[\mathbf{S}_2 = \mathbf{s}] = \frac{\exp \boldsymbol{\lambda}'\mathbf{s}}{\sum_{\mathbf{s} \in \mathcal{Q}} \exp \boldsymbol{\lambda}'\mathbf{s}} = \frac{\prod_{k \in U} \omega_k^{s_k}}{\sum_{\mathbf{s} \in \mathcal{Q}} \prod_{k \in U} \omega_k^{s_k}}, \text{ for all } \mathbf{s} \in \mathcal{Q},$$

we get

$$\Pr[\mathbf{S}_1 + \mathbf{S}_2 = \mathbf{s}] = \sum_{\mathbf{v} \in \mathcal{S}_1} \Pr[\mathbf{S}_2 = \mathbf{s} - \mathbf{v}] \Pr[\mathbf{S}_1 = \mathbf{v}]$$

$$= \sum_{\ell \in U} s_\ell q_\ell \Pr[\mathbf{S} = \mathbf{s} - \mathbf{a}_\ell],$$

for all $\mathbf{s} \in \{\mathbf{r} + \mathbf{v}, \text{ for all } \mathbf{r} \in \mathcal{Q}, \mathbf{v} \in \mathcal{S}_1\}$, where $\mathbf{a}_\ell \in \mathcal{S}_1$ such that $a_{\ell\ell} = 1$ and $a_{\ell k} = 0$ when $k \neq \ell$. By conditioning on $\mathcal{S}_n$, we obtain

$$\Pr[\mathbf{S}_1 + \mathbf{S}_2 = \mathbf{s} | (\mathbf{S}_1 + \mathbf{S}_2) \in \mathcal{S}_n] = C_n \sum_{\ell \in U} \pi_\ell s_\ell \prod_{\substack{k \in U \\ k \neq \ell}} \omega_k^{s_k},$$

which is the Sampford sampling design.

Multinomial rejective Sampford's procedure

If $\mathcal{Q} = \mathcal{R}_n$, we obtain a simple implementation by means of a multinomial rejective procedure that is presented in Algorithm 7.3. This procedure can be very slow especially when the sample size is large with respect to the population size.

Algorithm 7.3 Multinomial rejective Sampford's procedure

1. Select n units with replacement, the first drawing being made with probabilities $q_k = \pi_k/n$ and all the subsequent ones with probabilities proportional to $\pi_k/(1-\pi_k)$.
2. If any unit is selected multiple times, reject the sample and restart the procedure.

Poisson rejective Sampford's procedure

If $\mathcal{Q} = \mathcal{S}$, we obtain a simple implementation by means of a Poisson rejective procedure that is presented in Algorithm 7.4.

Algorithm 7.4 Poisson rejective Sampford's procedure

1. Select the first unit with equal probability, $q_k = \pi_k/n$.
2. Select a sample by means of a Poisson sampling design (with replacement) with inclusion probabilities

$$\widetilde{\pi}_k = \frac{c\pi_k}{1 - (1 - c)\pi_k}, \text{ for any } c \in \mathbb{R}_+.$$

(Of course, one can choose $c = 1$, which gives $\widetilde{\pi}_k = \pi_k$.)
3. If any unit is selected twice or if the sample size is not equal to n, reject the sample and restart the procedure.

CPS rejective Sampford's procedure

If $\mathcal{Q} = \mathcal{S}_n$, we obtain an implementation by means of a CPS rejective procedure. Traat et al. (2004) have suggested the ingenious implementation presented in Algorithm 7.5.

Algorithm 7.5 CPS rejective Sampford's procedure

1. Select the first unit with equal probability, $q_k = \pi_k/n$. Let i be the selected unit.
2. Exchange the places of unit i and 1 in the list.
3. Perform a CPS design of size $n-1$ by means of sequential procedure 5.8, page 92, with parameter $\lambda_k = \log(\omega_k) = \log[\pi_k/(1 - \pi_k)]$.
4. If unit 1 (former i) is selected twice, reject the sample and restart the procedure.

7.5 Variance Approximation and Estimation

7.5.1 Approximation of the Variance

Each one of the unequal probability sampling methods presented in Chapters 5 through 7 has particular joint inclusion probabilities. Their variances of the estimator can theoretically be estimated by the Horvitz-Thompson estimator (see Expression 2.18, page 28) or the Sen-Yates-Grundy estimator of the variance (see Expression 2.19, page 28). In practice, the use of joint inclusion probabilities is often unrealistic because they are difficult to compute and n^2 terms must be summed to compute the estimate.

Ideally, a practical estimator should be computed by a simple sum over the sample, and the computation of the joint inclusion probabilities should be avoided. Such an estimator exists for the multinomial sampling design (see Expression (5.5), page 72), but multinomial sampling is not satisfactory. Indeed, sampling with replacement can always be improved (see Section 5.4.2, page 72). It is, however, possible to conclusively approximate the variance for several sampling methods by using a quite simple estimator. On this matter, Berger (1998a) showed that if the sampling design does not diverge too much from the exponential sampling design, it is possible to construct an approximation.

Numerous publications (Hájek, 1981; Hartley and Chakrabarty, 1967; Deville, 1993; Brewer, 1999; Aires, 2000b; Brewer, 2001; Brewer and Donadio, 2003; Matei and Tillé, 2006) are devoted to the derivation of an approximation of the variance. A simple reasoning can provide an approximation of variance for exponential designs without replacement: Result 26, page 89 shows that a CPS design $p(\mathbf{s})$ is obtained by conditioning a Poisson design $\widetilde{p}(\mathbf{s})$ given that its sample size $\widetilde{n}_S$ is fixed. If $\mathrm{var}_{\mathrm{POISSON}}(.)$ denotes the variance and $\mathrm{cov}_{\mathrm{POISSON}}(.)$ the covariance under the Poisson sampling design $\widetilde{p}(\mathbf{s})$ and $\mathrm{var}(.)$ the variance under the design $p(.)$, we can write

$$\mathrm{var}\left(\widehat{Y}_{HT}\right) = \mathrm{var}_{\mathrm{POISSON}}\left(\widehat{Y}_{HT}|\widetilde{n}_S = n\right).$$

If we suppose that the pair $(\widehat{Y}_{HT}, \widetilde{n}_S)$ has a bivariate normal distribution (on this topic see Hájek, 1964; Berger, 1998a), we obtain

$$\mathrm{var}_{\mathrm{POISSON}}\left(\widehat{Y}_{HT}|\widetilde{n}_S = n\right) = \mathrm{var}_{\mathrm{POISSON}}\left[\widehat{Y}_{HT} + (n - \widetilde{n}_S)\beta\right],$$

where

$$\beta = \frac{\mathrm{cov}_{\mathrm{POISSON}}\left(\widetilde{n}_S, \widehat{Y}_{HT}\right)}{\mathrm{var}_{\mathrm{POISSON}}\left(\widetilde{n}_S\right)},$$

$$\mathrm{var}_{\mathrm{POISSON}}\left(\widetilde{n}_S\right) = \sum_{k \in U} \widetilde{\pi}_k(1 - \widetilde{\pi}_k),$$

and

$$\text{cov}_{\text{POISSON}}\left(\tilde{n}_S, \widehat{Y}_{HT}\right) = \sum_{k \in U} \tilde{\pi}_k(1 - \tilde{\pi}_k)\frac{y_k}{\pi_k}.$$

If $b_k = \tilde{\pi}_k(1 - \tilde{\pi}_k)$, we get a general approximation of the variance for an exponential sampling design (see Deville and Tillé, 2005; Tillé, 2001, p. 117):

$$\text{var}_{\text{APPROX}}\left[\widehat{Y}_{HT}\right] = \sum_{k \in U} b_k \left(\breve{y}_k - \breve{y}^*\right)^2, \tag{7.13}$$

where $\breve{y}_k = y_k/\pi_k$ and

$$\breve{y}^* = \beta = \frac{\sum_{\ell \in U} b_\ell y_\ell / \pi_\ell}{\sum_{\ell \in U} b_\ell}.$$

According to the values given to b_k, we obtain numerous variants of this approximation. Indeed, Expression (7.13) can also be written

$$\text{var}_{\text{APPROX}}\left(\widehat{Y}_{HT}\right) = \sum_{k \in U} \frac{y_k^2}{\pi_k^2}\left(b_k - \frac{b_k^2}{\sum_{\ell \in U} b_\ell}\right) - \frac{1}{\sum_{\ell \in U} b_\ell} \sum_{k \in U} \sum_{\substack{\ell \in U \\ \ell \neq k}} \frac{y_k y_\ell b_k b_\ell}{\pi_k \pi_\ell}$$

or

$$\text{var}_{\text{APPROX}}\left(\widehat{Y}_{HT}\right) = \breve{\mathbf{y}}' \boldsymbol{\Delta}_{\text{APPROX}} \breve{\mathbf{y}},$$

where

$$\boldsymbol{\Delta}_{\text{APPROX}} = \text{diag}(\mathbf{b}) - \frac{\mathbf{bb}'}{\sum_{k \in U} b_k},$$

and $\mathbf{b} = (b_1 \cdots b_N)$. Each variant of $\mathbf{b}$ thus allows approximating the matrix $\boldsymbol{\Delta}$, from which is possible to derive an approximation of $\boldsymbol{\Pi}$ by $\boldsymbol{\Pi}_{\text{APPROX}} = \boldsymbol{\Delta}_{\text{APPROX}} + \boldsymbol{\pi}\boldsymbol{\pi}'$.

Hájek approximation 1

The most common value for b_k has been proposed by Hájek (1981):

$$b_k = \frac{\pi_k(1 - \pi_k)N}{N - 1} \tag{7.14}$$

(on this topic see also Rosén, 1997; Tillé, 2001).

Example 24. With $N = 6, n = 3$, and $\boldsymbol{\pi} = (0.07 \ 0.17 \ 0.41 \ 0.61 \ 0.83 \ 0.91)'$, we obtain

$$\mathbf{b}_1 = (0.07812 \ 0.16932 \ 0.29028 \ 0.28548 \ 0.16932 \ 0.09828),$$

$$\mathbf{\Delta}_{\mathrm{APPROX1}} = \begin{pmatrix} 0.073 & -0.011 & -0.02 & -0.019 & -0.011 & -0.006 \\ -0.011 & 0.143 & -0.044 & -0.043 & -0.025 & -0.014 \\ -0.02 & -0.044 & 0.213 & -0.075 & -0.044 & -0.025 \\ -0.019 & -0.043 & -0.075 & 0.211 & -0.043 & -0.025 \\ -0.011 & -0.025 & -0.044 & -0.043 & 0.143 & -0.014 \\ -0.006 & -0.014 & -0.025 & -0.025 & -0.014 & 0.089 \end{pmatrix},$$

and

$$\mathbf{\Pi}_{\mathrm{APPROX1}} = \begin{pmatrix} 0.077 & 0 & 0.008 & 0.022 & 0.046 & 0.057 \\ 0 & 0.172 & 0.025 & 0.059 & 0.115 & 0.139 \\ 0.008 & 0.025 & 0.381 & 0.174 & 0.295 & 0.347 \\ 0.022 & 0.059 & 0.174 & 0.583 & 0.462 & 0.529 \\ 0.046 & 0.115 & 0.295 & 0.462 & 0.832 & 0.740 \\ 0.057 & 0.139 & 0.347 & 0.529 & 0.740 & 0.918 \end{pmatrix}. \tag{7.15}$$

Note that the diagonal of $\mathbf{\Pi}_{\mathrm{APPROX1}}$ is not equal to the inclusion probabilities.

Fixed-point approximation

The general approximation given in (7.13) can also be written:

$$\mathrm{var}_{\mathrm{APPROX}}\left(\widehat{Y}_{HT}\right) = \sum_{k \in U} \frac{y_k^2}{\pi_k^2}\left(b_k - \frac{b_k^2}{\sum_{\ell \in U} b_\ell}\right) - \frac{1}{\sum_{\ell \in U} b_\ell} \sum_{k \in U} \sum_{\substack{\ell \in U \\ \ell \neq k}} \frac{y_k y_\ell b_k b_\ell}{\pi_k \pi_\ell}.$$

The exact variance can be written

$$\mathrm{var}(\widehat{Y}_{HT}) = \sum_{k \in U} \frac{y_k^2}{\pi_k^2} \pi_k (1 - \pi_k) + \sum_{k \in U} \sum_{\substack{\ell \in U \\ \ell \neq k}} \frac{y_k y_\ell}{\pi_k \pi_\ell}(\pi_{k\ell} - \pi_k \pi_\ell).$$

Deville and Tillé (2005) proposed to compute b_k in such a way that the coefficient of the y_k^2's are exact, which amounts to solving the following equation system to find another approximation of b_k,

$$b_k - \frac{b_k^2}{\sum_{\ell \in U} b_\ell} = \pi_k(1 - \pi_k). \tag{7.16}$$

In Matei and Tillé (2006), a large set of simulations shows that this approximation is very accurate. Because the equation system (7.16) is not linear, the coefficients b_k can be obtained by the fixed-point technique, using the following recurrence equation until convergence:

$$b_k^{(i)} = \frac{\left[b_k^{(i-1)}\right]^2}{\sum_{l \in U} b_l^{(i-1)}} + \pi_k(1 - \pi_k), \tag{7.17}$$

for $i = 0, 1, 2, 3, \ldots$, and using the initialization:

$$b_k^{(0)} = \pi_k(1 - \pi_k)\frac{N}{N-1}, k \in U.$$

Unfortunately, Equation (7.16) has no solution unless (see Deville and Tillé, 2005)

$$\frac{\pi_k(1 - \pi_k)}{\sum_{\ell \in U} \pi_\ell(1 - \pi_\ell)} \leq \frac{1}{2}, \text{ for all } k \text{ in } U.$$

If the method is not convergent, one can use the following variant, which consists of using one iteration,

$$b_k^{(1)} = \pi_k(1 - \pi_k)\left[\frac{N\pi_k(1 - \pi_k)}{(N-1)\sum_{\ell \in U} \pi_\ell(1 - \pi_\ell)} + 1\right].$$

Example 25. With $N = 6, n = 3$, $\boldsymbol{\pi} = (0.07\ 0.17\ 0.41\ 0.61\ 0.83\ 0.91)'$, we obtain

$$\mathbf{b}_2 = (0.06922, 0.1643, 0.3431, 0.3335, 0.1643, 0.08866),$$

$$\boldsymbol{\Delta}_{\text{APPROX2}} = \begin{pmatrix} 0.065 & -0.009 & -0.019 & -0.019 & -0.009 & -0.004 \\ -0.009 & 0.141 & -0.047 & -0.046 & -0.022 & -0.012 \\ -0.019 & -0.047 & 0.242 & -0.097 & -0.047 & -0.025 \\ -0.019 & -0.046 & -0.097 & 0.238 & -0.046 & -0.024 \\ -0.009 & -0.022 & -0.047 & -0.046 & 0.141 & -0.012 \\ -0.004 & -0.012 & -0.025 & -0.024 & -0.012 & 0.082 \end{pmatrix}$$

and

$$\boldsymbol{\Pi}_{\text{APPROX2}} = \begin{pmatrix} 0.07 & 0.002 & 0.008 & 0.023 & 0.048 & 0.058 \\ 0.002 & 0.17 & 0.021 & 0.057 & 0.118 & 0.142 \\ 0.008 & 0.021 & 0.41 & 0.152 & 0.292 & 0.347 \\ 0.023 & 0.057 & 0.152 & 0.61 & 0.459 & 0.530 \\ 0.048 & 0.118 & 0.292 & 0.459 & 0.83 & 0.743 \\ 0.058 & 0.142 & 0.347 & 0.530 & 0.743 & 0.91 \end{pmatrix}. \quad (7.18)$$

Note that the diagonal of $\boldsymbol{\Pi}_{\text{APPROX2}}$ is now equal to the inclusion probabilities.

7.5.2 Estimators Based on Approximations

From the general approximation given in (7.13), it is possible to construct a class of estimators that depend only on first-order inclusion probabilities for all $k \in S$. A general variance estimator can be derived (see Deville and Tillé, 2005; Tillé, 2001, p. 117):

$$\widehat{\text{var}}(\widehat{Y}_{HT}) = \sum_{k \in U} S_k c_k \left(\breve{y}_k - \hat{y}^*\right)^2, \quad (7.19)$$

where $\breve{y}_k = y_k/\pi_k$ and

$$\hat{y}^* = \frac{\sum_{\ell \in U} S_\ell c_\ell y_\ell/\pi_\ell}{\sum_{\ell \in U} S_\ell c_\ell}.$$

A large set of values has been proposed for c_k.

Estimator 1

A simple value for c_k can be (Deville, 1993, see, for instance) :

$$c_k = (1 - \pi_k) \frac{n}{n-1}. \tag{7.20}$$

Estimator 2 of Deville

In the same manuscript, Deville (1993) suggested a more complex value (see also Deville, 1999):

$$c_k = (1 - \pi_k) \left\{ 1 - \sum_{k \in U} S_k \left[\frac{1 - \pi_k}{\sum_{\ell \in U} S_\ell (1 - \pi_\ell)} \right]^2 \right\}^{-1}.$$

Estimator based on the approximation

When the b_k are defined by solving Equation (7.16), the following coefficient can be constructed:

$$c_k = \frac{(N-1)n}{N(n-1)} \frac{b_k}{\pi_k},$$

which gives

$$\widehat{\text{var}}_1[\widehat{Y}_{HT}] = \frac{n(N-1)}{N(n-1)} \sum_{k \in U} S_k \frac{b_k}{\pi_k} (\breve{y}_k - \hat{\breve{y}}^*)^2, \tag{7.21}$$

where

$$\hat{\breve{y}}^* = \frac{\sum_{\ell \in U} S_\ell b_\ell y_\ell / \pi_\ell^2}{\sum_{\ell \in U} S_\ell b_\ell / \pi_\ell}.$$

In the case where the inclusion probabilities are equal, we obtain the variance of simple random sampling without replacement. Unfortunately, the inclusion probabilities must be known for the entire population in order to compute the c_k.

Fixed-point estimator

Deville and Tillé (2005) proposed to use the following development to derive a value for c_k. The estimator defined in Expression (7.19) can also be written as

$$\widehat{\text{var}}[\widehat{Y}_{HT}] = \sum_{k \in U} S_k \frac{y_k^2}{\pi_k^2} \left(c_k - \frac{c_k^2}{\sum_{\ell \in U} S_\ell c_\ell} \right) - \frac{1}{\sum_{\ell \in U} S_\ell c_\ell} \sum_{k \in U} S_k \sum_{\substack{\ell \in U S_\ell \\ \ell \neq k}} \frac{y_k y_\ell c_k c_\ell}{\pi_k \pi_\ell}. \tag{7.22}$$

Using Formula (2.18), we can look for c_k which satisfies the equation:

$$c_k - \frac{c_k^2}{\sum_{\ell \in U} S_\ell c_\ell} = (1 - \pi_k). \tag{7.23}$$

These coefficients can be obtained by the fixed-point technique, using the following recurrence equation until the convergence is fulfilled:

$$c_k^{(i)} = \frac{\left[c_k^{(i-1)}\right]^2}{\sum_{\ell \in U} S_\ell c_\ell^{(i-1)}} + (1 - \pi_k),$$

for $i = 0, 1, 2, 3, \ldots$ and using the initialization:

$$c_k^{(0)} = (1 - \pi_k) \frac{n}{n-1}, k \text{ such that } S_k > 0.$$

Unfortunately, Equation (7.23) has no solution unless the following inequality is satisfied (see Deville and Tillé, 2005):

$$\frac{1 - \pi_k}{\sum_{\ell \in U} S_\ell (1 - \pi_\ell)} \leq \frac{1}{2}, \text{ for all } k \text{ in } S.$$

If the method is not convergent, one can use the previous variant, which uses one iteration:

$$c_k^{(1)} = (1 - \pi_k) \left[\frac{n(1 - \pi_k)}{(n-1) \sum_{\ell \in U} S_\ell (1 - \pi_\ell)} + 1 \right].$$

It is difficult to conclude that a particular estimator is better than the other ones. Nevertheless, a large set of simulations (see Deville, 1993; Brewer and Donadio, 2003; Matei and Tillé, 2006; Brewer, 2002, Chapter 9) shows that the estimator given in Expression (7.20) gives very satisfactory results and is almost always better than the Horvitz-Thompson and the Sen-Yates-Grundy estimators.

7.6 Comparisons of Methods with Unequal Probabilities

There exist a large number of sampling methods with unequal probabilities and fixed sample size. When the same inclusion probabilities are used, none of the methods without replacement provides a better accuracy than the other ones, as shown in the following result.

Result 33. *Let $p_1(.)$ and $p_2(.)$ be two sampling designs without replacement with the same first-order inclusion probabilities $\pi_1, \ldots, \pi_N$. Their variance-covariance operators are denoted $\mathbf{\Delta}_1$ and $\mathbf{\Delta}_2$. If, for some $\mathbf{u} \in \mathbb{R}^N$,*

$$\mathbf{u}' \mathbf{\Delta}_1 \mathbf{u} < \mathbf{u}' \mathbf{\Delta}_2 \mathbf{u},$$

then there exists at least one $\mathbf{v} \in \mathbb{R}^N$ such that

$$\mathbf{v}' \mathbf{\Delta}_1 \mathbf{v} > \mathbf{v}' \mathbf{\Delta}_2 \mathbf{v}.$$

Proof. (by contradiction, as given by Lionel Qualité)
Suppose that for all $\mathbf{x} \in \mathbb{R}^N, \mathbf{x}'\boldsymbol{\Delta}_1\mathbf{x} \leq \mathbf{x}'\boldsymbol{\Delta}_2\mathbf{x}$, and that $\mathbf{u}'\boldsymbol{\Delta}_1\mathbf{u} < \mathbf{u}'\boldsymbol{\Delta}_2\mathbf{u}$.
It follows that $\mathbf{x}'(\boldsymbol{\Delta}_2 - \boldsymbol{\Delta}_1)\mathbf{x} \geq 0$. Matrix $\boldsymbol{\Delta}_2 - \boldsymbol{\Delta}_1$ is thus positive semi-definite. Now, $p_1(.)$ and $p_2(.)$ have the same inclusion probabilities, which implies that $\boldsymbol{\Delta}_1$ and $\boldsymbol{\Delta}_2$ have the same trace. Thus, $\text{trace}(\boldsymbol{\Delta}_1 - \boldsymbol{\Delta}_2) = 0$. Because $\boldsymbol{\Delta}_2 - \boldsymbol{\Delta}_1$ is positive semi-definite, all the eigenvalues of $\boldsymbol{\Delta}_1 - \boldsymbol{\Delta}_2$ are null and $\mathbf{x}'(\boldsymbol{\Delta}_2 - \boldsymbol{\Delta}_1)\mathbf{x} = 0$, which is contradictory to $\mathbf{u}'\boldsymbol{\Delta}_1\mathbf{u} < \mathbf{u}'\boldsymbol{\Delta}_2\mathbf{u}$. $\square$

Result 33 is quite disappointing because it does not allow choosing a sampling design. The variances of different sampling designs, however, can be compared with each other. An interesting indicator consists of computing the largest possible deviation (LPD) of variance between two sampling designs:

$$\text{LPD}(2|1) = \max_{\mathbf{u}|\mathbf{u}'\boldsymbol{\pi}\neq 0} \frac{\mathbf{u}'\boldsymbol{\Delta}_2\mathbf{u}}{\mathbf{u}'\boldsymbol{\Delta}_1\mathbf{u}} - 1.$$

Note that the LPD is equal to 0 if $\boldsymbol{\Delta}_1$ is equal to $\boldsymbol{\Delta}_2$, but the LPD is not symmetrical; that is, $\text{LPD}(1|2) \neq \text{LPD}(2|1)$. Practically $\text{LPD}(2|1)$ is the largest eigenvalue of matrix $\boldsymbol{\Delta}_1^+\boldsymbol{\Delta}_2$, where $\boldsymbol{\Delta}_1^+$ is the Moore-Penrose inverse of $\boldsymbol{\Delta}_1$.

Example 26. The LPD has been computed to compare several sampling designs for which it is possible to compute the matrix of joint inclusion probabilities:

- The CPS design (see Expression (5.26), page 86),
- Sampford's method (see Expression (7.12), page 134),
- Tillé's eliminatory method (see Expression (6.6), page 116),
- The random systematic method (see Expression (7.3), page 128).

These four matrices of inclusion probabilities are next compared to

- The Hájek approximation (Approx 1) (see Expression (7.15), page 139),
- The fixed-point approximation (Approx 2) (see Expression (7.18), page 140).

Table 7.4 contains the LPD computed among the four sampling designs and the two approximations.

Figure 7.3 contains a multidimensional scaling of the matrix of the LPD given in Table 7.4. First Table 7.4 and its transpose are added in order to obtain a symmetric matrix. Next, the optimal Euclidian representation in $\mathbb{R}^2$ is obtained by multidimensional scaling.

Although the inclusion probabilities are very uneven, Table 7.4 shows that the CPS design and the Sampford design are very close and are well approximated by the fixed-point approximation (Approx 2). For instance, the use of fixed-point approximation (Approx 2) in place of the true variance for a CPS provides a maximum overestimation of the variance equal to 8.5% and a maximum underestimation of 14.5%. For the standard deviation, the maximum overestimation is 4.1% and the maximum underestimation is 7.0%.

Table 7.4. Largest possible deviation for the CPS, Sampford, Tillé, and random systematic designs and two approximations of Δ for Example 26; for instance, column 1 and row 2 contain LPD(CPS | Sampford)

	CPS	Sampford	Tillé	Systematic	Approx 1	Approx 2
CPS	0	0.032	0.157	0.385	0.134	0.085
Sampford	0.040	0	0.122	0.439	0.125	0.052
Tillé	0.374	0.322	0	0.876	0.228	0.204
Systematic	0.694	0.747	0.958	0	0.773	0.837
Approx 1	0.240	0.214	0.190	0.583	0	0.175
Approx 2	0.142	0.099	0.069	0.574	0.123	0

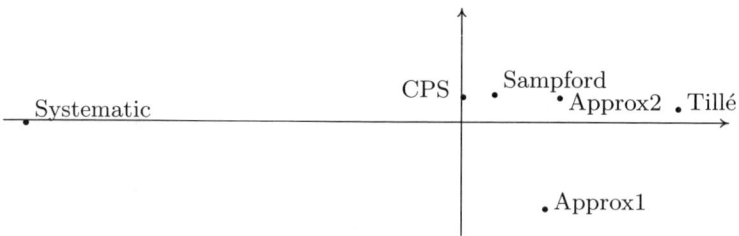

Fig. 7.3. Multidimensional scaling (MDS) of the symmetrized matrix of the LPD from Example 26

Example 27. With another vector of inclusion probabilities

$$\boldsymbol{\pi} = (0.15 \quad 0.3 \quad 0.35 \quad 0.45 \quad 0.5 \quad 0.55 \quad 0.6 \quad 0.65 \quad 0.7 \quad 0.75)',$$

$N = 10$ and $n = 5$, Table 7.5 contains the matrix of the LPD. The vector $\boldsymbol{\pi}$ is less uneven and the LPD's are thus still reduced.

Table 7.5. Largest possible deviation for the CPS, Sampford, Tillé and random systematic designs and two approximations of Δ for Example 27

	CPS	Sampford	Tillé	Systematic	Approx 1	Approx 2
CPS	0	0.003	0.111	0.166	0.048	0.007
Sampford	0.010	0	0.108	0.177	0.047	0.004
Tillé	0.306	0.293	0	0.511	0.294	0.277
Systematic	0.141	0.144	0.151	0	0.151	0.147
Approx 1	0.023	0.019	0.079	0.189	0	0.018
Approx 2	0.024	0.013	0.105	0.191	0.046	0

Figure 7.4 contains a multidimensional scaling of the symmetrized matrix of the LPD given in Table 7.5. Next, the optimal Euclidian representation

in $\mathbb{R}^2$ is obtained by multidimensional scaling. Although the distances are smaller, the configuration is very similar to the previous example.

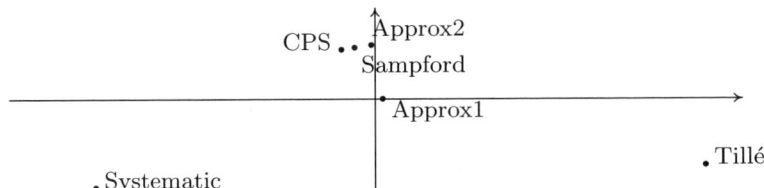

Fig. 7.4. Multidimensional scaling (MDS) of the symmetrized matrix of the LPD for Example 27

7.7 Choice of a Method of Sampling with Unequal Probabilities

Which sampling method with unequal probabilities to choose? We advocate the CPS design or the Sampford procedure. The CPS method is exponential and is linked with the Poisson sampling design and the multinomial sampling. Sampford's method is obviously a good alternative that is simple to implement. Obviously, random systematic sampling is already very much used due to its extreme simplicity, but the variance will be less well approximated and estimated. Nevertheless, it is now possible to select balanced samples with unequal probabilities (see Chapter 8), which provides more accurate estimators.

8

Balanced Sampling

8.1 Introduction

A balanced sampling design has the important property that the Horvitz-Thompson estimators of the totals for a set of auxiliary variables are equal to the totals we want to estimate. Therefore, the variances of all the variables of interest are reduced, depending on their correlations with the auxiliary variables. Yates (1949) had already insisted on the idea of respecting the means of known variables in probabilistic samples because the variance is then reduced. Yates (1946) and Neyman (1934) described methods of balanced sampling limited to one variable and to equal inclusion probabilities.

Royall and Herson (1973a) stressed the importance of balancing a sample for protecting inference against a misspecified model. They called this idea 'robustness'. Since no method existed for achieving a multivariate balanced sample, they proposed the use of simple random sampling, which is mean-balanced with large samples. More recently, several partial solutions were proposed by Deville et al. (1988), Deville (1992), Ardilly (1991), and Hedayat and Majumdar (1995). Valliant et al. (2000) surveyed some existing methods.

In this chapter, after surveying the most usual methods for equal inclusion probabilities, we present the cube method proposed by Deville and Tillé (2004, 2005) that allows selecting balanced samples with equal or unequal inclusion probabilities.

8.2 The Problem of Balanced Sampling

8.2.1 Definition of Balanced Sampling

The aim is always to estimate the total

$$Y = \sum_{k \in U} y_k.$$

Suppose also that the vectors of values

$$\mathbf{x}_k = (x_{k1} \; \cdots \; x_{kj} \; \cdots \; x_{kp})'$$

taken by p auxiliary variables are known for all units of the population. The vector of population totals of the balancing variables is thus also known

$$\mathbf{X} = \sum_{k \in U} \mathbf{x}_k,$$

and can be estimated by

$$\widehat{\mathbf{X}}_{HT} = \sum_{k \in U} \frac{\mathbf{x}_k S_k}{\pi_k}.$$

The aim is to construct a balanced sampling design, defined as follows.

Definition 53. *A sampling design $p(\mathbf{s})$ is said to be balanced with respect to the auxiliary variables $x_1, \ldots, x_p$, if and only if it satisfies the balancing equations given by*

$$\widehat{\mathbf{X}}_{HT} = \mathbf{X}, \qquad\qquad (8.1)$$

which can also be written

$$\sum_{k \in U} \frac{x_{kj} S_k}{\pi_k} = \sum_{k \in U} x_{kj},$$

for all $s \in \mathcal{S}$ such that $p(\mathbf{s}) > 0$ and for all $j = 1, \ldots, p$; or in other words

$$\mathrm{var}\left(\widehat{\mathbf{X}}_{HT}\right) = 0.$$

Balanced sampling can thus be viewed as a restriction on the support. Indeed, only the samples that satisfy the balancing equations have a strictly positive probability; that is, the support is

$$\mathcal{Q} = \left\{ \mathbf{s} \in \mathcal{S} \; \middle| \; \sum_{k \in U} \frac{\mathbf{x}_k s_k}{\pi_k} = \mathbf{X} \right\}.$$

8.2.2 Particular Cases of Balanced Sampling

Some particular cases of balanced sampling are well-known. Consider the following two particular cases.

Particular case 1: Fixed sample size

A sampling design of fixed sample size n is balanced on the variable $x_k = \pi_k, k \in U$, because

$$\sum_{k \in U} \frac{x_k S_k}{\pi_k} = \sum_{k \in U} S_k = \sum_{k \in U} \pi_k = n.$$

Remark 12. In sampling with unequal probabilities, when all the inclusion probabilities are different, the Horvitz-Thompson estimator of the population size N is generally random:

$$\widehat{N}_{HT} = \sum_{k \in U} \frac{S_k}{\pi_k}.$$

When the population size is known before selecting the sample, it could be important to select a sample such that

$$\sum_{k \in U} \frac{S_k}{\pi_k} = N. \tag{8.2}$$

Equation (8.2) is a balancing equation, in which the balancing variable is $x_k = 1, k \in U$. We show that balancing equation (8.2) can be satisfied by means of the cube method.

Particular case 2: stratification

Suppose that the population can be split into H nonoverlapping groups (or strata) denoted $U_h, h = 1, \ldots, H$, of sizes $N_h, h = 1, \ldots, H$. A sampling design is said to be stratified if a sample is selected in each stratum with simple random sampling with fixed sample sizes $n_1, \ldots, n_h \ldots, n_H$, the samples being selected independently in each stratum.

  Now, let $\delta_{k1}, \ldots, \delta_{kH}$, where

$$\delta_{kh} = \begin{cases} 1 & \text{if } k \in U_h \\ 0 & \text{if } k \notin U_h. \end{cases}$$

A stratified design is balanced on the variables $\delta_{k1}, \ldots, \delta_{kH}$; that is,

$$\sum_{k \in U} \frac{S_k \delta_{kh}}{\pi_k} = \sum_{k \in U} \delta_{kh} = N_h.$$

for $h = 1, \ldots, H$. In other words, the Horvitz-Thompson estimators of the N_h's are exactly equal to N_h. Stratification is thus a particular case of balanced sampling.

8.2.3 The Rounding Problem

In most cases, the balancing equations (8.1) cannot be exactly satisfied, as Example 28 shows.

Example 28. Suppose that $N = 10, n = 7, \pi_k = 7/10, k \in U$, and the only auxiliary variable is $x_k = k, k \in U$. Then, a balanced sample satisfies

$$\sum_{k \in U} \frac{k S_k}{\pi_k} = \sum_{k \in U} k,$$

so that $\sum_{k \in U} k S_k$ has to be equal to $55 \times 7/10 = 38.5$, which is impossible because 38.5 is not an integer. The problem arises because $1/\pi_k$ is not an integer and the population size is small.

The rounding problem also exists for simple random sampling and stratification. If the sum of the inclusion probabilities is not an integer, it is not possible to select a simple random sample with fixed sample size. In proportional stratification, all the inclusion probabilities are equal. In this case, it is generally impossible to define integer-value sample sizes in the strata in such a way that the inclusion probabilities are equal for all strata.

The rounding problem exists for almost all practical cases of balanced sampling. For this reason, our objective is to construct a sampling design, that satisfies the balancing equations (8.1) exactly if possible and to find the best approximation if this cannot be achieved. This rounding problem becomes negligible when the expected sample size is large.

8.3 Procedures for Equal Probabilities

8.3.1 Neyman's Method

Neyman (1934, p. 572) (see also Thionet, 1953, pp. 203-207) was probably the first author to propose a sampling procedure based on the use of small strata:

> Thus, we see that the method of purposive selection consists, (a) in dividing the population of districts into second-order strata according to values of y and x and (b) in selecting randomly from each stratum a definite number of districts. The number of samplings is determined by the condition of maintenance of the weighted average of the y.

8.3.2 Yates's Procedure

Yates (1946, p. 16) also proposed a method based on the random substitution of units.

> A random sample is first selected (stratified or other factors if required). Further members are then selected by the same random process, the first member being compared with the first member of the original sample, the second with the second member and so on, the new member being substituted for the original member if balancing is thereby improved.

The Neyman and Yates procedures are thus limited to equal probability sampling with only one balancing variable.

8.3.3 Deville, Grosbras, and Roth Procedure

Deville et al. (1988) proposed a method for the selection of balanced samples on several variables with equal inclusion probabilities. This method is based on the construction of groups that are constructed according to the means of all auxiliary variables. If p is not too large, we can construct 2^p groups according to the possible vector of signs:

$$\text{sign}\left(x_{k1} - \overline{X}_1 \cdots x_{kj} - \overline{X}_j \cdots x_{kp} - \overline{X}_p\right),$$

where

$$\overline{X}_j = \frac{1}{N} \sum_{k \in U} x_{kj}, \quad j = 1, \ldots, p.$$

The objective of Algorithm 8.1 is to select a sample balanced on

$$\left(\overline{X}_1 \cdots \overline{X}_j \cdots \overline{X}_p\right).$$

Unfortunately, Algorithm 8.1 provides an equal probability sampling design. The inclusion probabilities are approximately satisfied. Moreover, the 2^p groups can be difficult to compute when p is large, for instance, $2^{20} = 1\ 048\ 576$.

8.4 Cube Representation of Balanced Samples

8.4.1 Constraint Subspace and Cube of Samples

As we have seen in Section 2.4, in sampling without replacement, each vector $\mathbf{s}$ is a vertex of an N-cube and the number of possible samples is the number of vertices of an N-cube. A sampling design with inclusion probabilities $\pi_k, k \in U$, consists of assigning a probability $p(\mathbf{s})$ to each vertex of the N-cube such that

$$\text{E}(\mathbf{S}) = \sum_{s \in \mathcal{S}} p(\mathbf{s})\mathbf{s} = \boldsymbol{\pi},$$

where $\boldsymbol{\pi} = (\pi_k)$ is the vector of inclusion probabilities. Geometrically, a sampling design consists of expressing vector $\boldsymbol{\pi}$ as a convex linear combination of the vertices of the N-cube. A sampling algorithm can thus be viewed as a "random" way of reaching a vertex of the N-cube from a vector $\boldsymbol{\pi}$ in such a way that the balancing equations (8.1) are satisfied.

The balancing equations (8.1) can also be written

$$\begin{cases} \sum_{k \in U} \check{\mathbf{x}}_k s_k = \sum_{k \in U} \check{\mathbf{x}}_k \pi_k \\ s_k \in \{0, 1\}, k \in U, \end{cases} \tag{8.3}$$

where $\check{\mathbf{x}}_k = \mathbf{x}_k / \pi_k, k \in U$, and s_k equals 1 if unit k is in the sample and 0 otherwise. The first equation of (8.3) with given $\check{\mathbf{x}}_k$ and coordinates s_k defines

Algorithm 8.1 Deville, Grosbras, and Roth balanced procedure for equal inclusion probabilities

1. *Initialization.* First, assume by convention that the population mean $\overline{\mathbf{X}} = (\bar{X}_1 \cdots \bar{X}_j \cdots \bar{X}_p)'$ is a null vector of $\mathbb{R}^p$. Split the population into G groups, where $G \geq p$ and p is the number of balancing variables. Let $\overline{\mathbf{X}}_g, g = 1, \ldots, G$ be the vectors of $\mathbb{R}^p$ containing the means of the gth group. The groups must be constructed in such a way that the sizes N_g of the groups are nearly equal. Next, compute $n_g = nN_g/N$. Note that n_g is not necessarily integer.

2. *Rounding the sample group size.* The n_g's are rounded by means of an unequal probability algorithm (see Section 5.6 and Chapters 6 and 7) with fixed sample size and without replacement. Select a sample denoted $(I_1 \cdots I_g \cdots I_G)'$ from population $\{1, \ldots, g, \ldots, G\}$ with unequal inclusion probabilities $(n_1 - \lfloor n_1 \rfloor \cdots n_g - \lfloor n_g \rfloor \cdots n_G - \lfloor n_G \rfloor)'$, where $\lfloor n_g \rfloor$ is the greatest integer number less than or equal to n_g. The rounded group sizes are $m_g = \lfloor n_g \rfloor + I_g, g = 1, \ldots, G$.

3. *Computation of the sample mean.* Set $t = 1$ and $\check{\mathbf{x}}(1) = \sum_g \mu_g \bar{\mathbf{x}}_g$.

4. *Computation of $\boldsymbol{\nu}$.* Search a vector $\boldsymbol{\nu} = (\nu_1 \cdots \nu_g \cdots \nu_G)'$ by solving the following program

$$\text{minimize} \sum_{g=1}^{G} \frac{\nu_g^2}{N_g},$$

   subject to

$$\sum_{g=1}^{G} \nu_g = 0 \quad \text{and} \quad \sum_{g=1}^{G} \nu_g \overline{\mathbf{X}}_g = -n\check{\mathbf{x}}(t).$$

5. *Rounding $\boldsymbol{\nu}$.* The ν_g will be rounded by means of an unequal probability algorithm (see Section 5.6 and Chapters 6 and 7) with fixed sample size and without replacement. Select a sample denoted $(I_1 \cdots I_g \cdots I_G)$ from population $\{1, \ldots, g, \ldots, G\}$ with unequal inclusion probabilities $(\nu_1 - \lfloor \nu_1 \rfloor \cdots \nu_g - \lfloor \nu_g \rfloor \cdots \nu_G - \lfloor \nu_G \rfloor)$. The rounded ν_g are $\mu_g = \lfloor n_g \rfloor + I_g, g = 1, \ldots, G$.

6. *Stopping rule.* If $\sum_g |\mu_g| \geq A$, retain the sample and stop; otherwise set $t = t+1$ and compute $A = \sum_g |\mu_g|$.

7. *Updating the sample.* If $\mu_g > 0$ draw μ_g units from the sample by simple random sampling and drop them. If $\mu_g < 0$ draw $|\mu_g|$ new units from group g by simple random sampling and add them to the sample.

8. *Computation of the sample mean.* Compute the new sample mean $\check{\mathbf{x}}(t)$ and GOTO STEP 4.

an affine subspace Q in $\mathbb{R}^N$ of dimension $N - p$. Note that $Q = \boldsymbol{\pi} + \text{Ker } \mathbf{A}$, where

$$\text{Ker } \mathbf{A} = \{\mathbf{v} \in \mathbb{R}^N | \mathbf{A}\mathbf{v} = \mathbf{0}\}$$

is the kernel of the $p \times N$ matrix $\mathbf{A}$ given by $\mathbf{A} = (\check{\mathbf{x}}_1 \cdots \check{\mathbf{x}}_k \cdots \check{\mathbf{x}}_N)$. The main idea in obtaining a balanced sample is to choose a vertex of the N-cube that remains in the subspace Q or near Q if that is not possible.

If $C = [0, 1]^N$ denotes the N-cube in $\mathbb{R}^N$ whose vertices are the samples of U, the intersection between C and Q is nonempty because $\boldsymbol{\pi}$ is in the interior of C and belongs to Q. The intersection between an N-cube and a subspace defines a convex polytope $K = C \cap Q$ which has a dimension $N - p$ because it is the intersection of an N-cube and a plane of dimension $N - p$ that has a point in the interior of C.

8.4.2 Geometrical Representation of the Rounding Problem

The fact that the balancing equation system is exactly, approximately, or sometimes satisfied depends on the values of $\check{\mathbf{x}}_k$ and $\mathbf{X}$.

Definition 54. *Let D be a convex polytope. A vertex, or extremal point, of D is defined as a point that cannot be written as a convex linear combination of the other points of D. The set of all the vertices of D is denoted by $\mathrm{Ext}(D)$.*

Definition 55. *A sample $\mathbf{s}$ is said to be exactly balanced if $\mathbf{s} \in \mathrm{Ext}(C) \cap Q$.*

Note that a necessary condition for finding an exactly balanced sample is that $\mathrm{Ext}(C) \cap Q \neq \emptyset$.

Definition 56. *A balancing equation system is*

(i) Exactly satisfied if $\mathrm{Ext}(C) \cap Q = \mathrm{Ext}(C \cap Q)$,
(ii) Approximately satisfied if $\mathrm{Ext}(C) \cap Q = \emptyset$,
(iii) Sometimes satisfied if $\mathrm{Ext}(C) \cap Q \neq \mathrm{Ext}(C \cap Q)$ and $\mathrm{Ext}(C) \cap Q \neq \emptyset$.

A balancing system can thus be exactly satisfied if the extremal point of the intersection of the cube and of the subspace Q are also vertices of the cube. This case is rather exceptional. However, the following result shows that the rounding problem concerns only a limited number of units.

Result 34. *If $\mathbf{r} = [r_k]$ is a vertex of $K = C \cap Q$ then*

$$\mathrm{card}\{k | 0 < r_k < 1\} \leq p,$$

where p is the number of auxiliary variables and $\mathrm{card}(U)$ denotes the cardinality of U.

Proof. Let $\mathbf{A}^*$ be the matrix $\mathbf{A} = (\check{\mathbf{x}}_1 \cdots \check{\mathbf{x}}_k \cdots \check{\mathbf{x}}_N)$ restricted to the non-integer components of vector $\mathbf{r}$; that is, restricted to $U^* = \{k | 0 < r_k < 1\}$. If $q = \mathrm{card}(U^*) > p$, then $\mathrm{Ker}\, \mathbf{A}^*$ has dimension $q - p > 0$, and $\mathbf{r}$ is not an extreme point of K. $\qquad\square$

The following three examples show that the rounding problem can be viewed geometrically. Indeed, the balancing equations cannot be exactly satisfied when the vertices of K are not vertices of C; that is, when $q > 0$.

Example 29. In Figure 8.1, a sampling design in a population of size $N = 3$ is considered. The only constraint consists of fixing the sample size $n = 2$, and thus $p = 1$ and $x_k = \pi_k, k \in U$. The inclusion probabilities satisfy $\pi_1 + \pi_2 + \pi_3 = 2$, so that the balancing equation is exactly satisfied. We thus have $\mathbf{A} = (1, 1, 1)$, $\mathbf{A}\boldsymbol{\pi} = 2$, $Q = \{g_1, g_2, g_3 \in \mathbb{R} | g_1 + g_2 + g_3 = 2\}$, $\mathrm{Ext}(Q \cap C) = \{(1, 1, 0), (1, 0, 1), (0, 1, 1)\}$, and $\mathrm{Ext}(C) \cap Q = \{(1, 1, 0), (1, 0, 1), (0, 1, 1)\}$.

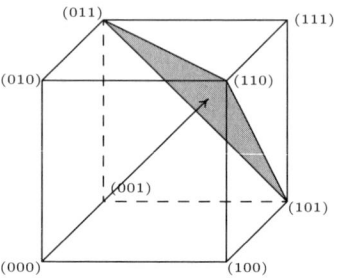

Fig. 8.1. Fixed size constraint of Example 29: all the vertices of K are vertices of the cube

Example 30. Figure 8.2 exemplifies the case where the constraint subspace does not pass through any vertices of the cube. The inclusion probabilities are $\pi_1 = \pi_2 = \pi_3 = 0.5$. The only constraint, $p = 1$, is given by the auxiliary variable $x_1 = 0, x_2 = 6\pi_2$, and $x_3 = 4\pi_3$. It is then impossible to satisfy exactly the balancing equation; the balancing equation is always approximately satisfied. We thus have $\mathbf{A} = (0, 6, 4)$, $\mathbf{A}\boldsymbol{\pi} = 5$, $Q = \{g_1, g_2, g_3 \in \mathbb{R} | 6g_2 + 4g_3 = 5\}$, $\mathrm{Ext}(Q \cap C) = \{(0, 5/6, 0), (0, 1/6, 1), (1, 5/6, 0), (1, 1/6, 0)\}$, and $\mathrm{Ext}(C) \cap Q = \emptyset$.

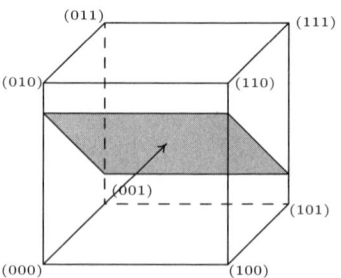

Fig. 8.2. Example 30: none of vertices of K are vertices of the cube

Example 31. Figure 8.3 exemplifies the case where the constraint subspace passes through two vertices of the cube but one vertex of the intersection is not a vertex of the cube. The inclusion probabilities are $\pi_1 = \pi_2 = \pi_3 = 0.8$. The only constraint, $p = 1$, is given by the auxiliary variable $x_1 = \pi_1, x_2 = 3\pi_2$ and $x_3 = \pi_3$. The balancing equation is only sometimes satisfied. In this case, there exist balanced samples, but there does not exist an exactly balanced sampling design for the given inclusion probabilities. In other words, although exact balanced samples exist, one must accept selecting only approximately balanced samples in order to satisfy the given inclusion probabilities. We thus have $\mathbf{A} = (1, 3, 1)$, $\mathbf{A}\boldsymbol{\pi} = 4$, $Q = \{g_1, g_2, g_3 \in \mathbb{R} | g_1 + 3g_2 + g_3 = 4\}$, $\text{Ext}(Q \cap C) = \{(1, 1, 0), (0, 1, 1), (1, 2/3, 1)\}$, $\text{Ext}(C) \cap Q = \{(1, 1, 0), (0, 1, 1)\}$.

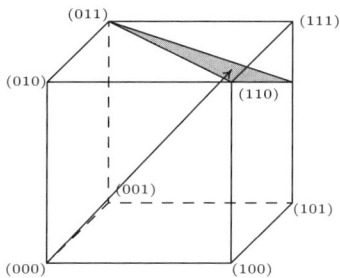

Fig. 8.3. Example 31: some vertices of K are vertices of the cube and others are not

8.5 Enumerative Algorithm

A first way to find a balanced sampling design is to define a cost function $\text{Cost}(\mathbf{s})$ for all possible samples of $\mathcal{S}$. The choice of the cost function is an arbitrary decision that depends on the priorities of the survey manager. The cost must be such that

- $\text{Cost}(\mathbf{s}) > 0$, for all $\mathbf{s} \in \mathcal{S}$.
- $\text{Cost}(\mathbf{s}) = 0$, if $\mathbf{s}$ is balanced.

A simple cost could be defined by the sum of squares function

$$\text{Cost}_1(\mathbf{s}) = \sum_{j=1}^{p} \frac{\left[\widehat{X}_{jHT}(\mathbf{s}) - X_j\right]^2}{X_j^2},$$

where $\widehat{X}_{jHT}(\mathbf{s})$ is the value taken by $\widehat{X}_{jHT}$ on sample $\mathbf{s}$. Instead, we might take

$$\text{Cost}_2(\mathbf{s}) = (\mathbf{s} - \boldsymbol{\pi}^*)' \mathbf{A}' \left(\mathbf{A}\mathbf{A}'\right)^{-1} \mathbf{A}(\mathbf{s} - \boldsymbol{\pi}^*),$$

where $\mathbf{A} = (\check{\mathbf{x}}_1 \cdots \check{\mathbf{x}}_k \cdots \check{\mathbf{x}}_N)$. The choice of $\text{Cost}_2(.)$ has a natural interpretation as a distance in $\mathbb{R}^N$, as shown by the following result.

Result 35. *The square of the distance between a sample $\mathbf{s}$ and its Euclidean projection onto the constraint subspace is given by*

$$\text{Cost}_2(\mathbf{s}) = (\mathbf{s} - \boldsymbol{\pi})'\mathbf{A}'\left(\mathbf{A}\mathbf{A}'\right)^{-1}\mathbf{A}(\mathbf{s} - \boldsymbol{\pi}). \tag{8.4}$$

Proof. The projection of a sample $\mathbf{s}$ onto the constraint subspace is

$$\mathbf{s} - \mathbf{A}'\left(\mathbf{A}\mathbf{A}'\right)^{-1}\mathbf{A}(\mathbf{s} - \boldsymbol{\pi}).$$

The square of the Euclidean distance between $\mathbf{s}$ and its projection is thus

$$(\mathbf{s}-\boldsymbol{\pi})'\mathbf{A}'\left(\mathbf{A}\mathbf{A}'\right)^{-1}\mathbf{A}(\mathbf{s}-\boldsymbol{\pi}) = (\mathbf{s}-\boldsymbol{\pi}^*+\boldsymbol{\pi}^*-\boldsymbol{\pi})'\mathbf{A}'\left(\mathbf{A}\mathbf{A}'\right)^{-1}\mathbf{A}(\mathbf{s}-\boldsymbol{\pi}^*+\boldsymbol{\pi}^*-\boldsymbol{\pi})$$

and, because $\mathbf{A}(\boldsymbol{\pi} - \boldsymbol{\pi}^*) = \mathbf{0}$, (8.4) follows directly. □

The sample can then be selected by means of Algorithm 8.2.

Algorithm 8.2 Balanced sampling by linear programming

Solve

$$\min_{p(\mathbf{s})} \sum_{\mathbf{s} \in \mathcal{S}} p(\mathbf{s})\text{Cost}(\mathbf{s}), \tag{8.5}$$

subject to

$$\sum_{\mathbf{s} \in \mathcal{S}} p(\mathbf{s}) = 1,$$

$$\sum_{\mathbf{s} \in \mathcal{S}} \mathbf{s}p(\mathbf{s}) = \boldsymbol{\pi},$$

and

$$p(\mathbf{s}) \geq 0, \text{ for all } \mathbf{s} \in \mathcal{S}.$$

Next, select the sample by means of the enumerative method (see Algorithm 3.1, page 32).

A linear program always gives particular sampling designs defined in the following way.

Definition 57. *Let $p(.)$ be a sampling design on a population U with inclusion probabilities π_k, and let $\mathcal{B} = \{\mathbf{s}|p(\mathbf{s}) > 0\}$. A sampling design $p(.)$ is said to be defined on a minimum support if and only if there does not exist a subset $\mathcal{B}_0 \subset \mathcal{B}$ such that $\mathcal{B}_0 \neq \mathcal{B}$ and*

$$\sum_{\mathbf{s} \in \mathcal{B}_0} p_0(\mathbf{s})s_k = \pi_k, \quad k \in U, \tag{8.6}$$

has a solution in $p_0(\mathbf{s})$.

Wynn (1977) studied sampling designs defined on minimum supports. The minimum support design presented in Algorithm 6.4, page 103, for selecting unequal probability sampling designs gives sampling designs on minimum supports.

Result 36. *The linear program (8.5) has at least one solution defined on a minimum support.*

The proof follows directly from the fundamental theorem of linear programming.

Example 32. Suppose that $N = 8$ and $n = 4$. The first balancing variable is proportional to π_k, which corresponds to a fixed size sampling design. The second variable is $x_k = k, k = 1, \ldots, N$. The inclusion probabilities are equal. Thus, we have

$$\pi = \begin{pmatrix} 0.5 \\ 0.5 \\ 0.5 \\ 0.5 \\ 0.5 \\ 0.5 \\ 0.5 \\ 0.5 \end{pmatrix}, \mathbf{X} = \begin{pmatrix} 1 & 1 \\ 1 & 2 \\ 1 & 3 \\ 1 & 4 \\ 1 & 5 \\ 1 & 6 \\ 1 & 7 \\ 1 & 8 \end{pmatrix}, \mathbf{A}' = \begin{pmatrix} 2 & 2 \\ 2 & 4 \\ 2 & 6 \\ 2 & 8 \\ 2 & 10 \\ 2 & 12 \\ 2 & 14 \\ 2 & 16 \end{pmatrix},$$

and

$$\mathbf{A}\pi = (8, 36)'.$$

The total of the second variable is thus $X = 36$. Table 8.1 contains the sampling design selected by linear programming, the cost $\text{Cost}_2(\mathbf{s})$ of each possible sample, and their Horvitz-Thompson estimator. Eight samples are exactly

Table 8.1. Sample cost and balanced sampling design for Example 32

s	$\text{Cost}_2(\mathbf{s})$	$\widehat{X}_{HT}$	$p(\mathbf{s})$	$\widehat{X}_{HT}p(\mathbf{s})$
(0, 0, 1, 1, 1, 1, 0, 0)	0	36	0.5	18
(1, 1, 0, 0, 0, 0, 1, 1)	0	36	0.5	18
			1	36

balanced and are

$$(0,0,1,1,1,1,0,0), (0,1,0,1,1,0,1,0), (0,1,1,0,0,1,1,0), (0,1,1,0,1,0,0,1),$$

$$(1,0,0,1,0,1,1,0), (1,0,0,1,1,0,0,1), (1,0,1,0,0,1,0,1), (1,1,0,0,0,0,1,1).$$

Nevertheless, the linear program proposes a solution where only two samples have a strictly positive probability of being selected. Other balanced sampling designs are possible because the solution of a linear program is not necessarily unique.

The enumerative methods have several drawbacks. Firstly, their implementation is limited by the combinatory explosion of the number of samples. Enumerative methods are thus limited to a population size $N \leq 20$. Moreover, the design obtained by a linear program has a very small support, which can be problematic to estimate the variance or to construct a confidence interval.

Example 33. Suppose that $N = 8$ and $n = 4$. The first balancing variable is π_k, which corresponds to a fixed size sampling design. The second variable is $x_k = 1, k = 1, \ldots, N$, which implies

$$\sum_{k \in U} \frac{S_k}{\pi_k} = \sum_{k \in U} 1 = N.$$

The inclusion probabilities are unequal because $\pi_k = k/9$. Thus, we obtain

$$\pi = \begin{pmatrix} 1/9 \\ 2/9 \\ 3/9 \\ 4/9 \\ 5/9 \\ 6/9 \\ 7/9 \\ 8/9 \end{pmatrix}, \mathbf{X} = \begin{pmatrix} 1/9 \ 1 \\ 2/9 \ 1 \\ 3/9 \ 1 \\ 4/9 \ 1 \\ 5/9 \ 1 \\ 6/9 \ 1 \\ 7/9 \ 1 \\ 8/9 \ 1 \end{pmatrix}, \mathbf{A}' = \begin{pmatrix} 1 \ \ 9 \\ 1 \ 9/2 \\ 1 \ 9/3 \\ 1 \ 9/4 \\ 1 \ 9/5 \\ 1 \ 9/6 \\ 1 \ 9/7 \\ 1 \ 9/8 \end{pmatrix},$$

and

$$\mathbf{A}\pi = (4, 8)'.$$

The total of the second variable is thus $X = 8$. Since the first balancing variable corresponds to a fixed size constraint, we can restrict the linear program to $\mathcal{S}_n$. Table 8.2 contains the selected sampling design, the cost $\mathrm{Cost}_2(\mathbf{s})$ of each possible sample, and their Horvitz-Thompson estimator. The sampling

Table 8.2. Sample cost and balanced sampling design for Example 33

s	$\mathrm{Cost}_2(\mathbf{s})$	$\widehat{X}_{HT}$	$p(\mathbf{s})$	$\widehat{X}_{HT}p(\mathbf{s})$	$(\widehat{X}_{HT} - X)^2 p(\mathbf{s})$
(0, 0, 0, 1, 1, 1, 0, 1)	0.03588	6.675	0.222	1.4833	0.3901
(0, 0, 0, 1, 1, 1, 1, 0)	0.02770	6.836	0.111	0.7595	0.1506
(0, 0, 1, 0, 1, 0, 1, 1)	0.01273	7.211	0.222	1.6024	0.1384
(0, 0, 1, 1, 0, 0, 1, 1)	0.00235	7.661	0.111	0.8512	0.0128
(0, 1, 0, 0, 0, 1, 1, 1)	0.00345	8.411	0.222	1.8690	0.0375
(1, 0, 0, 0, 0, 1, 1, 1)	0.49285	12.911	0.111	1.4345	2.6794
			1	8	3.406

design contains only six samples. Note that the most balanced sample, which is $(0, 0, 1, 1, 0, 1, 1, 0)$ and has a cost equal to 0.00003, is not in this sampling design.

Remark 13. Example 33 shows that the sampling design which minimizes the average cost does not necessarily allocate a nonnull probability to the most balanced sample.

Table 3.1, page 32 shows the limits of the enumerative methods. Even if we limit ourselves to samples of fixed size, the enumerative methods cannot deal with populations larger than 30 units. In order to select a balanced sample, the cube method provides the necessary shortcut to avoid the enumeration of samples.

8.6 The Cube Method

The cube method is composed of two phases, called the flight phase and the landing phase. In the flight phase, the constraints are always exactly satisfied. The objective is to randomly round off to 0 or 1 almost all of the inclusion probabilities; that is, to randomly select a vertex of $K = Q \cap C$. The landing phase consists of managing as well as possible the fact that the balancing equations (8.1) cannot always be exactly satisfied.

8.6.1 The Flight Phase

The aim of the flight phase is to randomly choose a vertex of

$$K = \{[0, 1]^N \cap Q\},$$

where $Q = \pi + \text{Ker } \mathbf{A}$, and $\mathbf{A} = (\check{\mathbf{x}}_1 \ \cdots \ \check{\mathbf{x}}_k \ \cdots \ \check{\mathbf{x}}_N)$, in such a way that the inclusion probabilities $\pi_k, k \in U$, and the balancing equations (8.1) are exactly satisfied. Note that by Result 34, a vertex of K has at most p noninteger values. The landing phase is necessary only if the attained vertex of K is not a vertex of C and consists of relaxing the constraints (8.1) as minimally as possible in order to select a sample; that is, a vertex of C.

The general algorithm for completing the flight phase is to use a balancing martingale.

Definition 58. *A discrete time stochastic process* $\pi(t) = [\pi_k(t)], t = 0, 1, \ldots$ *in* $\mathbb{R}^N$ *is said to be a balancing martingale for a vector of inclusion probabilities* π *and the auxiliary variables* $x_1, \ldots, x_p$ *if*

(i) $\pi(0) = \pi$,
(ii) $\text{E} \{\pi(t) | \pi(t-1), \ldots, \pi(0)\} = \pi(t-1), t = 1, 2, \ldots,$
(iii) $\pi(t) \in K = \{[0, 1]^N \cap (\pi + \text{Ker } \mathbf{A})\}$, *where* $\mathbf{A}$ *is the* $p \times N$ *matrix given by* $\mathbf{A} = (\mathbf{x}_1/\pi_1 \ \cdots \ \mathbf{x}_k/\pi_k \ \cdots \ \mathbf{x}_N/\pi_N)$.

A balancing martingale therefore satisfies that $\boldsymbol{\pi}(t-1)$ is the mean of the following possible values of $\boldsymbol{\pi}(t)$.

Result 37. *If $\boldsymbol{\pi}(t)$ is a balancing martingale, then we have the following:*

(*i*) $\mathrm{E}\left[\boldsymbol{\pi}(t)\right] = \mathrm{E}\left[\boldsymbol{\pi}(t-1)\right] = \cdots = \mathrm{E}\left[\boldsymbol{\pi}(0)\right] = \boldsymbol{\pi};$

(*ii*) $\displaystyle\sum_{k\in U} \check{\mathbf{x}}_k \pi_k(t) = \sum_{k\in U} \check{\mathbf{x}}_k \pi_k = \mathbf{X}, t = 0, 1, 2, \ldots;$

(*iii*) *When the balancing martingale reaches a face of C, it does not leave it.*

Proof. Part (*i*) is obvious. Part (*ii*) results from the fact that $\boldsymbol{\pi}(t) \in K$. To prove (*iii*), note that $\boldsymbol{\pi}(t-1)$ belongs to a face. It is the mean of the possible values of $\boldsymbol{\pi}(t)$ that therefore must also belong to this face. $\qquad\square$

Part (*iii*) of Result 37 directly implies that (*i*) if $\pi_k(t) = 0$, then $\pi_k(t+h) = 0, h \geq 0$; (*ii*) if $\pi_k(t) = 1$, then $\pi_k(t + h) = 1, h \geq 0$; and (*iii*) the vertices of K are absorbing states.

The practical problem is to find a method that rapidly reaches a vertex. Algorithm 8.3 allows us to attain a vertex of K in at most N steps.

Algorithm 8.3 General balanced procedure: Flight phase

INITIALIZE $\boldsymbol{\pi}(0) = \boldsymbol{\pi}$.

FOR $t = 0, \ldots, T$, and until it is no longer possible to carry out STEP 1, DO

1. Generate any vector $\mathbf{u}(t) = [u_k(t)] \neq 0$, random or not, such that $\mathbf{u}(t)$ is in the kernel of matrix $\mathbf{A}$, and $u_k(t) = 0$ if $\pi_k(t)$ is an integer number.
2. Compute $\lambda_1^*(t)$ and $\lambda_2^*(t)$, the largest values of $\lambda_1(t)$ and $\lambda_2(t)$ such that $0 \leq \boldsymbol{\pi}(t) + \lambda_1(t)\mathbf{u}(t) \leq 1$, and $0 \leq \boldsymbol{\pi}(t) - \lambda_2(t)\mathbf{u}(t) \leq 1$. Note that $\lambda_1(t) > 0$ and $\lambda_2(t) > 0$.
3. Select

$$\boldsymbol{\pi}(t + 1) = \begin{cases} \boldsymbol{\pi}(t) + \lambda_1^*(t)\mathbf{u}(t) & \text{with probability } q_1(t) \\ \boldsymbol{\pi}(t) - \lambda_2^*(t)\mathbf{u}(t) & \text{with probability } q_2(t), \end{cases} \qquad (8.7)$$

where $q_1(t) = \lambda_2^*(t)/\{\lambda_1^*(t) + \lambda_2^*(t)\}$ and $q_2(t) = \lambda_1^*(t)[\lambda_1^*(t) + \lambda_2^*(t)]$.

ENDFOR.

Figure 8.4 shows the geometric representation of the first step in a balancing martingale in the case of $N = 3$. The only constraint is the fixed sample size. Now, Algorithm 8.3 defines a balancing martingale. Clearly, $\boldsymbol{\pi}(0) = \boldsymbol{\pi}$. Also from Expression (8.7), we obtain $\mathrm{E}\left[\boldsymbol{\pi}(t)|\boldsymbol{\pi}(t-1), \ldots, \boldsymbol{\pi}(0)\right\} = \boldsymbol{\pi}(t-1), t = 1, 2, \ldots,$ because

$$\mathrm{E}\left[\boldsymbol{\pi}(t)|\boldsymbol{\pi}(t-1), \mathbf{u}(t)\right] = \boldsymbol{\pi}(t-1), \quad t = 1, 2, \ldots.$$

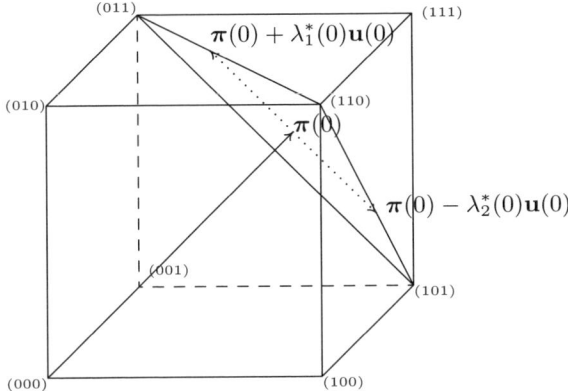

Fig. 8.4. Example for the first step of a balancing martingale in $\mathcal{S}_2$ and a population of size $N = 3$

Finally, because $\mathbf{u}(t)$ is in the kernel of $\mathbf{A}$, from (8.7) we obtain that $\boldsymbol{\pi}(t)$ always remains in $K = \left\{ [0, 1]^N \cap (\boldsymbol{\pi} + \text{Ker } \mathbf{A}) \right\}$.

At each step, at least one component of the process is rounded to 0 or 1. Thus, $\boldsymbol{\pi}(1)$ is on a face of the N-cube; that is, on a cube of dimension $N - 1$ at most, $\boldsymbol{\pi}(2)$ is on a cube of dimension $N - 2$ at most and so on. Let T be the time when the flight phase has stopped. The fact that step 1 is no longer possible shows that the balancing martingale has attained a vertex of K and thus by Result 34, page 153, that $\text{card}\{0 < \pi_k(T) < 1\} \leq p$.

8.6.2 Fast Implementation of the Flight Phase

Chauvet and Tillé (2006, 2005a,b) proposed an implementation of the flight phase that provides a very fast algorithm. In Algorithm 8.3, the search for a vector $\mathbf{u}$ in Ker $\mathbf{A}$ can be expensive. The basic idea is to use a submatrix $\mathbf{B}$ containing only $p + 1$ columns of $\mathbf{A}$. Note that the number of variables p is smaller than the population size N and that rank $\mathbf{B} \leq p$. The dimension of the kernel of $\mathbf{B}$ is thus larger than or equal to 1.

A vector $\mathbf{v}$ of Ker $\mathbf{B}$ can then be used to construct a vector $\mathbf{u}$ of Ker $\mathbf{A}$ by complementing $\mathbf{v}$ with zeros for the columns of $\mathbf{B}$ that are not in $\mathbf{A}$. With this idea, all the computations can be done only on $\mathbf{B}$. This method is described in Algorithm 8.4.

If $\widetilde{T}$ is the last step of the algorithm and $\widetilde{\boldsymbol{\pi}} = \boldsymbol{\pi}(\widetilde{T})$, then we have

1. $\text{E}(\widetilde{\boldsymbol{\pi}}) = \boldsymbol{\pi}$,
2. $\mathbf{A}\widetilde{\boldsymbol{\pi}} = \mathbf{A}\boldsymbol{\pi}$,
3. If $\widetilde{q} = \text{card}\{k | 0 < \widetilde{\pi}_k < 1\}$, then $\widetilde{q} \leq p$, where p is the number of auxiliary variables.

Algorithm 8.4 Fast algorithm for the flight phase

---

1. *Initialization*
    a) The units with inclusion probabilities equal to 0 or 1 are removed from the population before applying the algorithm in such a way that all the remaining units are such that $0 < \pi_k < 1$.
    b) The inclusion probabilities are loaded into vector $\boldsymbol{\pi}$.
    c) The vector $\boldsymbol{\psi}$ is made up of the first $p + 1$ elements of $\boldsymbol{\pi}$.
    d) A vector of ranks is created $\mathbf{r} = (1\ 2\ \cdots\ p\ p + 1)'$.
    e) Matrix $\mathbf{B}$ is made up of the first $p + 1$ columns of $\mathbf{A}$.
    f) INITIALIZE $k = p + 2$.
2. *Basic loop*
    a) A vector $\mathbf{u}$ is taken in the kernel of $\mathbf{B}$,
    b) Only $\boldsymbol{\psi}$ is modified (and not the vector $\boldsymbol{\pi}$) according to the basic technique. Compute λ_1^* and λ_2^*, the largest values of λ_1 and λ_2 such that $0 \le \boldsymbol{\psi} + \lambda_1 \mathbf{u} \le 1$, and $0 \le \boldsymbol{\psi} - \lambda_2 \mathbf{u} \le 1$. Note that $\lambda_1^* > 0$ and $\lambda_2^* > 0$.
    c) Select
    $$\boldsymbol{\psi} = \begin{cases} \boldsymbol{\psi} + \lambda_1^* \mathbf{u} & \text{with probability } q \\ \boldsymbol{\psi} - \lambda_2^* \mathbf{u} & \text{with probability } 1 - q, \end{cases}$$
    where $q = \lambda_2^*/(\lambda_1^* + \lambda_2^*)$.
    d) (*The units that correspond to $\psi(i)$ integer numbers are removed from $\mathbf{B}$ and are replaced by new units. The algorithm stops at the end of the file.*)
    FOR $i = 1, \ldots, p + 1$, DO
        IF $\psi(i) = 0$ or $\psi(i) = 1$ THEN
            IF $k \le N$ THEN $\begin{vmatrix} \boldsymbol{\pi}(\mathbf{r}(i)) = \psi(i); \\ \mathbf{r}(i) = k; \\ \psi(i) = \pi(k); \\ \text{FOR } j = 1, \ldots, p, \text{ DO } \mathbf{B}(i,j) = \mathbf{A}(k,j); \text{ ENDFOR}; \\ k = k + 1; \end{vmatrix}$
            ELSE GOTO STEP 3(a);
            ENDIF;
        ENDIF;
    ENDFOR.
    e) GOTO STEP 2(a).
3. *End of the first part of the flight phase*
    a) FOR $i = 1, \ldots, p + 1$, DO $\boldsymbol{\pi}(\mathbf{r}(i)) = \psi(i)$ ENDFOR.

In the case where some of the constraints can be satisfied exactly, the flight phase can be continued. Suppose that $\mathbf{C}$ is the matrix containing the columns of $\mathbf{A}$ that correspond to noninteger values of $\widetilde{\boldsymbol{\pi}}$, and $\boldsymbol{\phi}$ is the vector of noninteger values of $\widetilde{\boldsymbol{\pi}}$. If $\mathbf{C}$ is not full-rank, one or several steps of the general Algorithm 8.3 can still be applied to $\mathbf{C}$ and $\boldsymbol{\phi}$. A return to the general Algorithm 8.3 is thus necessary for the last steps.

The implementation of the fast algorithm is quite simple. Matrix $\mathbf{A}$ never has to be completely loaded in memory and thus remains in a file that can be read sequentially. For this reason, there does not exist any restriction on the population size because the execution time depends linearly on the population size. The search for a vector $\mathbf{u}$ in the submatrix $\mathbf{B}$ limits the choice of the direction $\mathbf{u}$. In most cases, only one direction is possible. In order to increase the randomness of the sampling design, the units can possibly be randomly mixed before applying Algorithm 8.4.

Another option consists of sorting the units by decreasing order of size. Indeed, from experience, with the general Algorithm 8.3, the rounding problem often concerns units with large size; that is, large inclusion probabilities or large values of $\check{x}_k$. With the fast Algorithm 8.4, the rounding problem often concerns the units that are at the end of the file. If the units are sorted by decreasing order of size, the fast algorithm will try to first balance the big units and the rounding problem will instead concern small units. Analogously, if we want to get an exact fixed weight of potatoes, it is more efficient to first put the large potatoes on the balance and to finish with the smallest potatoes. This popular idea can also be used to balance a sample, even if the problem is more complex because it is multivariate.

The idea of considering only one subset of the units already underlays in the moving stratification procedure (see Tillé, 1996b) that provides a smoothed effect of stratification. When $p = 1$ and the only auxiliary variable is $x_k = \pi_k$, then the problem of balanced sampling amounts to sampling with unequal probabilities and fixed sample size. In this case, $\mathbf{A} = (1 \ \cdots \ 1)$. At each step, matrix $\mathbf{B} = (1 \ 1)$ and $\mathbf{u} = (-1 \ 1)'$. Algorithm 8.4 can therefore be simplified dramatically and is identical to the pivotal method (see Algorithm 6.6, page 107).

8.6.3 The Landing Phase

At the end of the flight phase, the balancing martingale has reached a vertex of K, which is not necessarily a vertex of C. This vertex is denoted by $\boldsymbol{\pi}^* = [\pi_k^*] = \boldsymbol{\pi}(T)$. Let q be the number of noninteger components of this vertex. If $q = 0$, the algorithm is completed. If $q > 0$ some constraints cannot be exactly attained.

Landing phase by an enumerative algorithm

Definition 59. *A sample* $\mathbf{s}$ *is said to be compatible with a vector* $\boldsymbol{\pi}^*$ *if* $\pi_k^* = s_k$ *for all* k *such that* π_k^* *is an integer. Let* $\mathcal{C}(\boldsymbol{\pi}^*)$ *denote the set with* 2^q *elements of compatible samples with* $\boldsymbol{\pi}^*$.

It is clear that we can limit ourselves to finding a design with mean value $\boldsymbol{\pi}^*$ and whose support is included in $\mathcal{C}(\boldsymbol{\pi}^*)$.

The landing phase can be completed by an enumerative algorithm on subpopulation $\mathcal{C}(\boldsymbol{\pi}^*)$ as developed in Section 8.5. The following linear program provides a sampling design on $\mathcal{C}(\boldsymbol{\pi}^*)$.

$$\min_{p^*(.)} \sum_{\mathbf{s} \in \mathcal{C}(\boldsymbol{\pi}^*)} \text{Cost}(\mathbf{s}) p^*(\mathbf{s}), \tag{8.8}$$

subject to

$$\sum_{s \in \mathcal{C}(\boldsymbol{\pi}^*)} p^*(\mathbf{s}) = 1,$$

$$\sum_{s \in \mathcal{C}(\boldsymbol{\pi}^*)} \mathbf{s}p^*(\mathbf{s}) = \boldsymbol{\pi}^*,$$

$$0 \le p^*(\mathbf{s}) \le 1, \text{ for all } \mathbf{s} \in \mathcal{C}(\boldsymbol{\pi}^*).$$

Next, a sample is selected with sampling design $p^*(.)$. Because $q \le p$, this linear program no longer depends on the population size but only on the number of balancing variables. It is thus restricted to 2^q possible samples and, with a modern computer, can be applied without difficulty to a balancing problem with a score of auxiliary variables. If the inclusion probabilities are an auxiliary variable and the sum of the inclusion probabilities is integer, then the linear program can be applied only to

$$\mathcal{C}_n(\boldsymbol{\pi}^*) = \left\{ \mathbf{s} \in \mathcal{C}(\boldsymbol{\pi}^*) \middle| \sum_{k \in U} s_k = \sum_{k \in U} \pi_k = n \right\},$$

which dramatically limits the number of samples.

Landing phase by suppression of variables

If the number of balancing variables is too large for the linear program to be solved by a simplex algorithm, $q > 20$, then, at the end of the flight phase, an balancing variable can be straightforwardly suppressed. A constraint is thus relaxed, allowing a return to the flight phase until it is no longer possible to "move" within the constraint subspace. The constraints are thus successively relaxed. For this reason, it is necessary to order the balancing variables according to their importance so that the least important constraints are relaxed first. This naturally depends on the context of the survey.

8.6.4 Quality of Balancing

The rounding problem can arise with any balanced sampling design. For instance, in stratification, the rounding problem arises when the sum of the inclusion probabilities within the strata is not an integer number, which is almost always the case in proportional stratification or optimal stratification. In practice, the stratum sample sizes n_h are rounded either deterministically or randomly. Random rounding is used so as to satisfy in expectation the values of n_h. The purpose of random rounding is to respect the initial inclusion probabilities.

The cube method also uses a random rounding. In the particular case of stratification, it provides exactly the well-known method of random rounding of the sample sizes in the strata. With any variant of the landing phase, the deviation between the Horvitz-Thompson estimator and the total can be bounded because the rounding problem only depends on $q \le p$ values.

Result 38. *For any application of the cube method*

$$\left|\widehat{X}_{jHT} - X_j\right| \leq p \times \max_{k \in U} \left|\frac{x_{kj}}{\pi_k}\right|. \tag{8.9}$$

Proof.

$$\left|\widehat{X}_{jHT} - X_j\right| = \left|\sum_{k \in U} \frac{S_k x_{kj}}{\pi_k} - \sum_{k \in U} \frac{x_{kj}}{\pi_k} \pi_k\right| = \left|\sum_{k \in U} \frac{S_k x_{kj}}{\pi_k} - \sum_{k \in U} \frac{x_{kj}}{\pi_k} \pi_k^*\right|$$

$$= \left|\sum_{k \in U} \frac{S_k^* x_{kj}}{\pi_k} - \sum_{k \in U^*} \frac{x_{kj}}{\pi_k} \pi_k\right| \leq \sup_{F|\text{card}(F)=q} \sum_{k \in F} \left|\frac{x_{kj}}{\pi_k}\right|$$

$$\leq p \times \max_{k \in U} \left|\frac{x_{kj}}{\pi_k}\right|. \qquad \square$$

Result 39. *If the sum of the inclusion probabilities is integer and if the sampling design has a fixed sample size, then, for any application of the cube method,*

$$\left|\widehat{X}_{jHT} - X_j\right| \leq (p-1) \times \max_{k \in U} \left|\frac{x_{kj}}{\pi_k} - \frac{N\overline{X}_j}{n}\right|. \tag{8.10}$$

where

$$\overline{X}_j = \frac{1}{N} \sum_{k \in U} x_{kj}.$$

Proof. With the cube method, we can always satisfy the fixed sample size constraint when the sum of the inclusion probabilities is integer, which can be written

$$\sum_{k \in U} \frac{S_k \pi_k}{\pi_k} = \sum_{k \in U} \pi_k.$$

Thus, at the end of the flight phase, at most $p-1$ values of $\boldsymbol{\pi}^*$ are not integer. We obtain

$$\left|\widehat{X}_{jHT} - X_j\right| = \left|\sum_{k \in U} S_k \left(\frac{x_{kj} - c\pi_k}{\pi_k} - \sum_{k \in U} \frac{x_{kj} - c\pi_k}{\pi_k} \pi_k\right)\right|$$

$$\leq (p-1) \times \max_{k \in U} \left|\frac{x_{kj} - c\pi_k}{\pi_k}\right|,$$

for any $c \in \mathbb{R}$. If $c = N\overline{X}_j/n$, we get Result 39. $\qquad \square$

This bound is a conservative bound of the rounding error because we consider the worst case. Moreover, this bound is computed for a total and must be considered relative to the population size. Let $\alpha_k = \pi_k N/n, k \in U$. For almost all the usual sampling designs, we can admit that $1/\alpha_k$ is bounded when $n \to \infty$. Note that, for a fixed sample size

$$\frac{1}{N} \sum_{k \in U} \alpha_k = 1.$$

The bound for the estimation of the mean can thus be written:

$$\frac{|\widehat{X}_{jHT} - X_j|}{N} \le \frac{p}{n} \times \max_{k \in U} \left| \frac{x_{kj}}{\alpha_k} - \overline{X}_j \right| = O(p/n),$$

where $O(1/n)$ is a quantity that remains bounded when multiplied by n. The bound thus very quickly becomes negligible if the sample size is large with respect to the number of balancing variables.

For comparison note that with a single-stage sampling design such as simple random sampling or Bernoulli sampling, we have generally that

$$\frac{|\widehat{X}_{jHT} - X_j|}{N} = O_p(1/\sqrt{n})$$

(see, for example, Rosén, 1972; Isaki and Fuller, 1982).

Despite the overstatement of the bound, the gain obtained by balanced sampling is very important. The rate of convergence is much faster for balanced sampling than for a usual sampling design. In practice, except for the case of very small sample sizes, the rounding problem is thus negligible. Furthermore, the rounding problem also becomes problematic in stratification with very small sample sizes. In addition, this bound corresponds to the "worst case", whereas the landing phase is used to find the best one.

8.7 Application of the Cube Method to Particular Cases

8.7.1 Simple Random Sampling

Simple random sampling is a particular case of the cube method. Suppose that $\boldsymbol{\pi} = (n/N \;\cdots\; n/N \;\cdots\; n/N)$ and that the balancing variable is $x_k = n/N, k \in U$. We thus have $\mathbf{A} = (1 \;\cdots\; 1) \in \mathbb{R}^N$ and $\mathrm{Ker}\, \mathbf{A} = \left\{ \mathbf{v} \in \mathbb{R}^N | \sum_{k \in U} v_k = 0 \right\}.$

There are at least three ways to select a simple random sampling without replacement.

1. The first way consists of beginning the first step by using

$$\mathbf{u}(1) = \left(\frac{N-1}{N} \quad -\frac{1}{N} \;\cdots\; -\frac{1}{N} \right)'.$$

Then, $\lambda_1(1) = (N-n)/(N-1)$, $\lambda_2(1) = n/(N-1)$ and

$$\boldsymbol{\pi}(1) = \begin{cases} \left(1 \;\; \frac{n-1}{N-1} \;\cdots\; \frac{n-1}{N-1} \right)' & \text{with probability } q_1(1) \\ \left(0 \;\; \frac{n}{N-1} \;\cdots\; \frac{n}{N-1} \right)' & \text{with probability } q_2(1), \end{cases}$$

where $q_1(1) = \pi_1 = n/N$ and $q_2(1) = 1 - \pi_1 = (N - n)/N$. This first step corresponds exactly to the selection-rejection method for SRSWOR described in Algorithm 4.3, page 48.

2. The second way consists of sorting the data randomly before applying the cube method with any vectors $\mathbf{v}(t)$. Indeed, any choice of $\mathbf{v}(t)$ provides a fixed size sampling with inclusion probabilities $\pi_k = n/N$. A random sort applied before any equal probability sampling provides a simple random sampling (see Algorithm 4.5, page 50).

3. The third way consists of using a random vector $\mathbf{v} = (v_k)$, where the v_k are N independent identically distributed variables. Next, this vector is projected on Ker $\mathbf{A}$, which gives

$$u_k = v_k - \frac{1}{N} \sum_{k \in U} v_k.$$

Note that, for such v_k, it is obvious that a preliminary sorting of the data will not change the sampling design, which is thus a simple random sampling design.

An interesting problem occurs when the design has equal inclusion probabilities $\pi_k = \pi, k \in U$, such that $N\pi$ is not an integer number. If the only constraint implies a fixed sample size; that is, $x_k = 1, k \in U$, then the balancing equation can only be approximately satisfied. Nevertheless, the flight phase of the cube method works until $N\text{--}p = N-1$ elements of $\boldsymbol{\pi}^* = \boldsymbol{\pi}(N-1)$ are integer numbers. The landing phase consists of randomly deciding whether the last unit is drawn. The sample size is therefore equal to one of the two integer numbers nearest to $N\pi$.

8.7.2 Stratification

Stratification can be achieved by taking $x_{kh} = \delta_{kh} n_h / N_h, h = 1, \ldots, H$, where N_h is the size of stratum U_h, n_h is the sample stratum size and

$$\delta_{kh} = \begin{cases} 1 & \text{if } k \in U_h \\ 0 & \text{if } k \notin U_h. \end{cases}$$

In the first step, we use

$$u_k(1) = v_k(1) - \frac{1}{N_h} \sum_{\ell \in U_h} v_\ell(1), \quad k \in U_h.$$

The three strategies described in Section 8.7.1 for simple random sampling allow us to obtain directly a stratified random sample with simple random sampling within the strata. If the sums of the inclusion probabilities are not integer within the strata, the cube method randomly rounds the sample sizes of the strata so as to ensure that the given inclusion probabilities are exactly satisfied.

The interesting aspect of the cube method is that the stratification can be generalized to overlapping strata, which can be called "quota random design" or "cross-stratification". Suppose that two stratification variables are available, for example, in a business survey with "activity sector" and "region". The strata defined by the first variable are denoted by $U_{h.}, h = 1, \ldots, H$, and the strata defined by the second variable are denoted by $U_{.i}, i = 1, \ldots, K$. Next, define the $p = H + K$ balancing variables,

$$x_{kj} = \pi_k \times \begin{cases} I\left[k \in U_{j.}\right] & j = 1, \ldots, H \\ I\left[k \in U_{.(j-H)}\right] & j = H+1, \ldots, H+K, \end{cases}$$

where $I[.]$ is an indicator variable that takes the value 1 if the condition is true and 0 otherwise. The sample can now be selected directly by means of the cube method. The generalization to a multiple quota random design follows immediately. It can be shown (Deville and Tillé, 2000) that the quota random sampling can be exactly satisfied.

8.7.3 Unequal Probability Sampling with Fixed Sample Size

The unequal inclusion probability problem can be solved by means of the cube method. Suppose that the objective is to select a sample of fixed size n with inclusion probabilities $\pi_k, k \in U$, such that $\sum_{k \in U} \pi_k = n$. In this case, the only balancing variable is $x_k = \pi_k$. In order to satisfy this constraint, we must have

$$\mathbf{u} \in \operatorname{Ker} \mathbf{A} = \left\{ \mathbf{v} \in \mathbb{R}^N \;\middle|\; \sum_{k \in U} v_k = 0 \right\},$$

and thus

$$\sum_{k \in U} u_k(t) = 0. \tag{8.11}$$

Each choice, random or not, of vectors $\mathbf{u}(t)$ that satisfy (8.11) produces another unequal probability sampling method. Nearly all existing methods, except the rejective ones and the variations of systematic sampling, can easily be expressed by means of the cube method. In this case, the cube method is identical to the splitting method based on the choice of a direction described in Section 6.2.3, page 102.

The techniques of unequal probability sampling can always be improved. Indeed, in all the available unequal probability sampling methods with fixed sample size, the design is only balanced on a single variable. Nevertheless, two balancing variables are always available, namely, $x_{k1} = \pi_k, k \in U$, and $x_{k2} = 1, k \in U$. The first variable implies a fixed sample size and the second one implies that

$$\widehat{N}_{HT} = \sum_{k \in U} \frac{S_k}{\pi_k} = N.$$

In all methods, the sample is balanced on x_{k1} but not on x_{k2}. The balanced cube method allows us to satisfy both constraints approximately.

8.8 Variance Approximations in Balanced Sampling

8.8.1 Construction of an Approximation

The variance of the Horvitz-Thompson estimator is

$$\text{var}\left(\widehat{Y}_{HT}\right) = \sum_{k \in U} \sum_{\ell \in U} \frac{y_k y_k}{\pi_k \pi_\ell} \Delta_{k\ell}, \tag{8.12}$$

where

$$\Delta_{k\ell} = \begin{cases} \pi_{k\ell} - \pi_k \pi_\ell & \text{if } k \neq \ell \\ \pi_k(1 - \pi_k) & \text{if } k = \ell \end{cases}$$

and $\boldsymbol{\Delta} = [\Delta_{k\ell}]$. Matrix $\boldsymbol{\Delta}$ is called the variance-covariance operator. Thus, the variance of $\widehat{Y}_{HT}$ can theoretically be expressed and estimated by using the joint inclusion probabilities. Unfortunately, even in very simple cases like fixed sample sizes, the computation of $\boldsymbol{\Delta}$ is practically impossible.

Deville and Tillé (2005) have proposed approximating the variance by supposing that the balanced sampling can be viewed as a conditional Poisson sampling. A similar idea was also developed by Hájek (1981, p. 26, see also Section 7.5.1, page 139) for sampling with unequal probabilities and fixed sample size. In the case of Poisson sampling, which is a sampling design with no balancing variables, the variance of $\widehat{Y}_{HT}$ is easy to derive and can be estimated because only first-order inclusion probabilities are needed. If $\widetilde{\mathbf{S}}$ is the random sample selected by a Poisson sampling design and $\widetilde{\pi}_k, k \in U$, are the first-order inclusion probabilities of the Poisson design, then

$$\text{var}_{\text{POISSON}}\left(\widehat{Y}_{HT}\right) = \text{var}_{\text{POISSON}}\left(\sum_{k \in U} \frac{y_k}{\pi_k} \widetilde{S}_k\right) = \sum_{k \in U} \frac{y_k^2}{\pi_k^2} \widetilde{\pi}_k \left(1 - \widetilde{\pi}_k\right) = \mathbf{\check{y}}' \widetilde{\boldsymbol{\Delta}} \mathbf{\check{y}},$$
$$\tag{8.13}$$

where $\mathbf{\check{y}} = (\check{y}_1 \cdots \check{y}_k \cdots \check{y}_N)'$, $\check{y}_k = y_k/\pi_k$, and $\widetilde{\boldsymbol{\Delta}} = \text{diag}\left[\widetilde{\pi}_k(1 - \widetilde{\pi}_k)\right]$. Note that Expression (8.13) contains π_k, and $\widetilde{\pi}_k$ because the variance of the usual estimator (function of π_k's) is computed under Poisson sampling (function of $\widetilde{\pi}_k$'s). The π_k's are always known, but the $\widetilde{\pi}_k'$s are not necessarily known.

If we suppose that, through Poisson sampling, the vector $(\widehat{Y}_{HT} \ \widehat{\mathbf{X}}_{HT}')'$ has approximately a multinormal distribution, we obtain

$$\text{var}_{\text{POISSON}}\left(\widehat{Y}_{HT}|\widehat{\mathbf{X}}_{HT} = \mathbf{X}\right) \approx \text{var}_{\text{POISSON}}\left[\widehat{Y}_{HT} + \left(\mathbf{X} - \widehat{\mathbf{X}}_{HT}\right)' \boldsymbol{\beta}\right], \tag{8.14}$$

where

$$\boldsymbol{\beta} = \text{var}_{\text{POISSON}}\left(\widehat{\mathbf{X}}_{HT}\right)^{-1} \text{cov}_{\text{POISSON}}\left(\widehat{\mathbf{X}}_{HT}, \widehat{Y}_{HT}\right),$$

$$\text{var}_{\text{POISSON}}\left(\widehat{\mathbf{X}}_{HT}\right) = \sum_{k \in U} \frac{\mathbf{x}_k \mathbf{x}_k'}{\pi_k^2} \widetilde{\pi}_k \left(1 - \widetilde{\pi}_k\right),$$

and

$$\text{cov}_{\text{POISSON}}\left(\widehat{\mathbf{X}}_{HT}, \widehat{Y}_{HT}\right) = \sum_{k \in U} \frac{\mathbf{x}_k y_k}{\pi_k^2} \widetilde{\pi}_k \left(1 - \widetilde{\pi}_k\right).$$

Again $\text{var}_{\text{POISSON}}\left(\widehat{\mathbf{X}}_{HT}\right)$ and $\text{cov}_{\text{POISSON}}\left(\widehat{\mathbf{X}}_{HT}, \widehat{Y}_{HT}\right)$ are functions of π_k and $\widetilde{\pi}_k$ because we compute the variance of the usual Horvitz-Thompson estimator (function of π_k) under the Poisson sampling design (function of $\widetilde{\pi}_k$).

If

$$b_k = \widetilde{\pi}_k(1 - \widetilde{\pi}_k),$$

Expression (8.14) can also be written

$$\text{var}_{\text{APPROX}}\left(\widehat{Y}_{HT}\right) = \sum_{k \in U} b_k \left(\breve{y}_k - \breve{y}_k^*\right)^2, \tag{8.15}$$

where

$$\breve{y}_k^* = \breve{\mathbf{x}}_k' \left(\sum_{\ell \in U} b_\ell \breve{\mathbf{x}}_\ell \breve{\mathbf{x}}_\ell'\right)^{-1} \sum_{\ell \in U} b_\ell \breve{\mathbf{x}}_\ell \breve{y}_\ell = \left[\mathbf{A}'(\mathbf{A}\widetilde{\boldsymbol{\Delta}}\mathbf{A}')^{-1}\mathbf{A}\widetilde{\boldsymbol{\Delta}}\breve{\mathbf{y}}\right]_k$$

is a regression predictor of $\breve{y}_k$ for the suitable regression such that $\breve{y}_k - \breve{y}_k^*$ appears as a residual and $\mathbf{A} = (\breve{\mathbf{x}}_1 \cdots \breve{\mathbf{x}}_k \cdots \breve{\mathbf{x}}_N)$.

When $p = 1$ and the only balancing variable is $x_k = \pi_k$, then the problem of balanced sampling amounts to sampling with unequal probabilities and fixed sample size. The approximation of variance given in (8.15) is then equal to the approximation given in Expression (7.13), page 138. In this case, $\breve{y}_k^*$ is simply the mean of the $\breve{y}_k$'s with the weights b_k.

The weights b_k unfortunately are unknown because they depend on the $\widetilde{\pi}_k$'s, which are not exactly equal to the π_k. We thus propose to approximate the b_k. Note that Expression (8.15) can also be written

$$\text{var}_{\text{APPROX}}\left(\widehat{Y}_{HT}\right) = \breve{\mathbf{y}}' \boldsymbol{\Delta}_{\text{APPROX}} \breve{\mathbf{y}},$$

where $\boldsymbol{\Delta}_{\text{APPROX}} = \{\Delta_{appk\ell}\}$ is the approximated variance-covariance operator and

$$\Delta_{appk\ell} = \begin{cases} b_k - b_k \breve{\mathbf{x}}_k' \left(\sum_{i \in U} b_i \breve{\mathbf{x}}_i \breve{\mathbf{x}}_i'\right)^{-1} \breve{\mathbf{x}}_k b_k & k = \ell \\ b_k \breve{\mathbf{x}}_k' \left(\sum_{k \in U} b_i \breve{\mathbf{x}}_i \breve{\mathbf{x}}_i'\right)^{-1} \breve{\mathbf{x}}_\ell b_\ell & k \neq \ell. \end{cases} \tag{8.16}$$

Four variance approximations can be obtained by various definitions of the b_k's. These four definitions are denoted b_{k1}, b_{k2}, b_{k3}, and b_{k4} and permit the definition of four variance approximations denoted $V_\alpha, \alpha = 1, 2, 3, 4$, and four variance-covariance operators denoted $\boldsymbol{\Delta}_\alpha, \alpha = 1, 2, 3, 4$, by replacing in (8.15) and (8.16), b_k with, respectively, b_{k1}, b_{k2}, b_{k3}, and b_{k4}.

1. The first approximation is obtained by considering that at least for large sample sizes, $\pi_k \approx \widetilde{\pi}_k, k \in U$. Thus, we take $b_{k1} = \pi_k(1 - \pi_k)$.

2. The second approximation is obtained by applying a correction for the loss of degrees of freedom:

$$b_{k2} = \pi_k(1 - \pi_k)\frac{N}{N - p}.$$

This correction allows obtaining the exact expression for simple random sampling with fixed sample size.
3. The third approximation is derived from the fact that the diagonal elements of the variance-covariance operator $\mathbf{\Delta}$ of the true variance are always known and are equal to $\pi_k(1 - \pi_k)$. Thus, by defining

$$b_{k3} = \pi_k(1 - \pi_k)\frac{\text{trace } \mathbf{\Delta}}{\text{trace } \mathbf{\Delta}_1},$$

we can define the approximated variance-covariance operator $\mathbf{\Delta}_3$ that has the same trace as $\mathbf{\Delta}$.
4. Finally, the fourth approximation is derived from the fact that the diagonal elements $\mathbf{\Delta}_{\text{APPROX}}$ can be computed and are given in (8.16). The b_{k4} are constructed in such a way that $\Delta_{k\ell} = \Delta_{appk\ell}$, or in other words, that

$$\pi_k(1 - \pi_k) = b_k - b_k\check{\mathbf{x}}'_k\left(\sum_{k \in U} b_k\check{\mathbf{x}}_k\check{\mathbf{x}}'_k\right)^{-1}\check{\mathbf{x}}_k b_k, \quad k \in U. \qquad (8.17)$$

The determination of the b_{k4}'s then requires the resolution of the nonlinear equation system. This fourth approximation is the only one that provides the exact variance expression for stratification.

In Deville and Tillé (2005), a set of simulations is presented which shows that b_{k4} is indisputably the most accurate approximation.

8.8.2 Application of the Variance Approximation to Stratification

Suppose that the sampling design is stratified; that is, the population can be split into H nonoverlapping strata denoted $U_h, h = 1, \ldots, H$, of sizes $N_h, h = 1, \ldots, H$. The balancing variables are

$$x_{k1} = \delta_{k1}, \ldots, x_{kH} = \delta_{kH},$$

where

$$\delta_{kh} = \begin{cases} 1 & \text{if } k \in U_h \\ 0 & \text{if } k \notin U_h. \end{cases}$$

If a simple random sample is selected in each stratum with sizes $n_1, \ldots, n_H$, then the variance can be computed exactly:

$$\text{var}\left(\widehat{Y}_{HT}\right) = \sum_{h=1}^{H} N_h^2 \frac{N_h - n_h}{N_h}\frac{V_{yh}^2}{n_h},$$

where

$$V_{yh}^2 = \frac{1}{N_h - 1} \sum_{k \in U_h} \left(y_k - \overline{Y}_h\right)^2.$$

It is thus interesting to compute the four approximations given in Section 8.8.1 in this particular case.

1. The first approximation gives

$$b_{k1} = \pi_k(1 - \pi_k) = \frac{n_h}{N_h} \frac{N_h - n_h}{N_h}.$$

Next,

$$\check{y}_k^* = \frac{1}{n_h} \sum_{k \in U_h} y_k = \frac{N_h}{n_h} \overline{Y}_h,$$

and

$$\mathrm{var}_{\mathrm{APPROX1}}\left(\widehat{Y}_{HT}\right) = \sum_{h=1}^H N_h(N_h - 1)\frac{N_h - n_h}{N_h}\frac{V_{yh}^2}{n_h}.$$

2. The second approximation gives

$$b_{k2} = \frac{N}{N - H}\pi_k(1 - \pi_k) = \frac{N}{N - H}\frac{n_h}{N_h}\frac{N_h - n_h}{N_h}.$$

Next,

$$\check{y}_k^* = \frac{1}{n_h} \sum_{k \in U_h} y_k = \frac{N_h}{n_h} \overline{Y}_h,$$

and

$$\mathrm{var}_{\mathrm{APPROX2}}\left(\widehat{Y}_{HT}\right) = \frac{N}{N - H}\sum_{h=1}^H N_h(N_h - 1)\frac{N_h - n_h}{N_h}\frac{V_{yh}^2}{n_h}.$$

3. The third approximation gives

$$b_{k3} = \pi_k(1 - \pi_k)\frac{\mathrm{trace}\,\boldsymbol{\Delta}}{\mathrm{trace}\boldsymbol{\Delta}_1} = \frac{n_h}{N_h}\frac{N_h - n_h}{N_h}\frac{\sum_{h=1}^H \frac{n_h}{N_h}(N_h - n_h)}{\sum_{h=1}^H \frac{n_h}{N_h^2}(N_h - n_h)(N_h - 1)}.$$

Next,

$$\check{y}_k^* = \frac{1}{n_h} \sum_{k \in U_h} y_k = \frac{N_h}{n_h} \overline{Y}_h,$$

and

$$\mathrm{var}_{\mathrm{APPROX3}}\left(\widehat{Y}_{HT}\right)$$

$$= \left[\frac{\sum_{h=1}^H \frac{n_h}{N_h}(N_h - n_h)}{\sum_{h=1}^H \frac{n_h}{N_h^2}(N_h - n_h)(N_h - 1)}\right]\sum_{h=1}^H N_h(N_h - 1)\frac{N_h - n_h}{N_h}\frac{V_{yh}^2}{n_h}.$$

4. The fourth approximation gives

$$b_{k4} = \frac{n_h}{N_h - 1} \frac{N_h - n_h}{N_h}.$$

Next,

$$\breve{y}_k^* = \frac{1}{n_h} \sum_{k \in U_h} y_k = \frac{N_h}{n_h} \overline{Y}_h,$$

and

$$\mathrm{var}_{\mathrm{APPROX4}}\left(\widehat{Y}_{HT}\right) = \sum_{h=1}^{H} N_h^2 \frac{N_h - n_h}{N_h} \frac{V_{yh}^2}{n_h}.$$

Although the differences between the variance approximations $\mathrm{var}_{\mathrm{APPROX1}}$, $\mathrm{var}_{\mathrm{APPROX2}}$, $\mathrm{var}_{\mathrm{APPROX3}}$, and $\mathrm{var}_{\mathrm{APPROX4}}$ are small relative to the population size, $\mathrm{var}_{\mathrm{APPROX4}}$ is the only approximation that gives the exact variance of a stratified sampling design.

8.9 Variance Estimation

8.9.1 Construction of an Estimator of Variance

Because Expression (8.15) is a function of totals, we can substitute each total by its Horvitz-Thompson estimator (see, for instance, Deville, 1999) in order to obtain an estimator of (8.15). The resulting estimator for (8.15) is:

$$\widehat{\mathrm{var}}\left(\widehat{Y}_{HT}\right) = \sum_{k \in U} c_k S_k \left(\breve{y}_k - \hat{\breve{y}}_k^*\right)^2, \qquad (8.18)$$

where

$$\hat{\breve{y}}_k^* = \breve{\mathbf{x}}_k' \left(\sum_{\ell \in U} c_\ell S_\ell \breve{\mathbf{x}}_\ell \breve{\mathbf{x}}_\ell'\right)^{-1} \sum_{\ell \in U} c_\ell S_\ell \breve{\mathbf{x}}_\ell \breve{y}_\ell$$

is the estimator of the regression predictor of $\breve{y}_k$.

Note that (8.18) can also be written

$$\sum_{k \in U} \sum_{\ell \in U} \breve{y}_k D_{k\ell} \breve{y}_\ell S_k S_\ell,$$

where

$$D_{k\ell} = \begin{cases} c_k - c_k \breve{\mathbf{x}}_k' \left(\sum_{i \in U} c_i S_i \breve{\mathbf{x}}_i \breve{\mathbf{x}}_i'\right)^{-1} \breve{\mathbf{x}}_k c_k & k = \ell \\ c_k \breve{\mathbf{x}}_k' \left(\sum_{i \in U} c_i c_k \breve{\mathbf{x}}_i \breve{\mathbf{x}}_i'\right)^{-1} \breve{\mathbf{x}}_\ell c_\ell & k \neq \ell. \end{cases}$$

The five definitions of the c_k's are denoted $c_{k1}, c_{k2}, c_{k3}, c_{k4}$, and c_{k5} and allow defining five variance estimators by replacing c_k in Expression (8.18) by, respectively, $c_{k1}, c_{k2}, c_{k3}, c_{k4}$, and c_{k5}.

1. The first estimator is obtained by taking $c_{k1} = (1 - \pi_k)$.
2. The second estimator is obtained by applying a correction for the loss of degrees of freedom:

$$c_{k2} = (1 - \pi_k) \frac{n}{n - p}.$$

This correction for the loss of degrees of freedom gives the unbiased estimator in simple random sampling with fixed sample size.
3. The third estimator is derived from the fact that the diagonal elements of the true matrix $\Delta_{k\ell}/\pi_{k\ell}$ are always known and are equal to $1 - \pi_k$. Thus, we can use

$$c_{k3} = (1 - \pi_k) \frac{\sum_{k \in U} (1 - \pi_k) S_k}{\sum_{k \in U} D_{kk1} S_k},$$

where D_{kk1} is obtained by plugging c_{k1} in D_{kk}.
4. The fourth estimator can be derived from b_{k4} obtained by solving the equation system (8.17).

$$c_{k4} = \frac{b_{k4}}{\pi_k} \frac{n}{n - p} \frac{N - p}{N}.$$

5. Finally, the fifth estimator is derived from the fact that the diagonal elements D_{kk} are known. The c_{k5}'s are constructed in such a way that

$$1 - \pi_k = D_{kk}, \quad k \in U, \tag{8.19}$$

or in other words that

$$1 - \pi_k = c_k - c_k \check{\mathbf{x}}_k' \left(\sum_{i \in U} c_i S_i \check{\mathbf{x}}_i \check{\mathbf{x}}_i' \right)^{-1} \check{\mathbf{x}}_k c_k, \quad k \in U.$$

A necessary condition of the existence of a solution for equation system (8.19) is that

$$\max_k \frac{1 - \pi_k}{\sum_{\ell \in U} S_i (1 - \pi_\ell)} < \frac{1}{2}.$$

The choice of the weights c_k is tricky. Although they are very similar, an evaluation by means of a set of simulations should still be run.

8.9.2 Application to Stratification of the Estimators of Variance

The case of stratification is interesting because the unbiased estimator of variance in a stratified sampling design (with a simple random sampling in each stratum) is known and is equal to

$$\widehat{\text{var}}\left(\widehat{Y}_{HT}\right) = \sum_{h=1}^{H} N_h^2 \frac{N_h - n_h}{N_h} \frac{v_{yh}^2}{n_h},$$

where

$$v_{yh}^2 = \frac{1}{n_h - 1} \sum_{k \in U_h} S_k \left(y_k - \widehat{\overline{Y}}_h \right)^2.$$

It is thus interesting to compute the five estimators in the stratification case.

1. The first estimator gives

$$c_{k1} = (1 - \pi_k) = \frac{N_h - n_h}{N_h}.$$

Next,

$$\hat{\dot{y}}_k^* = \frac{N_h}{n_h} \frac{1}{n_h} \sum_{k \in U_h} y_k S_k = \frac{N_h}{n_h} \widehat{\overline{Y}}_h,$$

and

$$\widehat{\mathrm{var}}_{\mathrm{APPROX1}} \left(\widehat{Y}_{HT} \right) = \sum_{h=1}^{H} N_h^2 \frac{N_h - n_h}{N_h} \frac{n_h - 1}{n_h} \frac{v_{yh}^2}{n_h}.$$

2. The second estimator gives

$$c_{k2} = \frac{n}{n - H} (1 - \pi_k) = \frac{n}{n - H} \frac{N_h - n_h}{N_h}.$$

Next,

$$\hat{\dot{y}}_k^* = \frac{N_h}{n_h} \frac{1}{n_h} \sum_{k \in U_h} y_k S_k = \frac{N_h}{n_h} \widehat{\overline{Y}}_h,$$

and

$$\widehat{\mathrm{var}}_{\mathrm{APPROX2}} \left(\widehat{Y}_{HT} \right) = \frac{n}{n - H} \sum_{h=1}^{H} N_h^2 \frac{N_h - n_h}{N_h} \frac{n_h - 1}{n_h} \frac{v_{yh}^2}{n_h}.$$

3. The third estimator gives

$$c_{k3} = \frac{N_h - n_h}{N_h} \frac{n - \sum_{h=1}^{H} \frac{n_h^2}{N_h}}{n - H - \sum_{h=1}^{H} \frac{n_h}{N_h}(n_h - 1)}.$$

Next,

$$\hat{\dot{y}}_k^* = \frac{N_h}{n_h} \frac{1}{n_h} \sum_{k \in U_h} y_k S_k = \frac{N_h}{n_h} \widehat{\overline{Y}}_h,$$

and

$$\widehat{\mathrm{var}}_{\mathrm{APPROX3}} \left(\widehat{Y}_{HT} \right)$$

$$= \frac{n - \sum_{h=1}^{H} \frac{n_h^2}{N_h}}{n - H - \sum_{h=1}^{H} \frac{n_h}{N_h}(n_h - 1)} \sum_{h=1}^{H} N_h^2 \frac{N_h - n_h}{N_h} \frac{n_h - 1}{n_h} \frac{v_{yh}^2}{n_h}.$$

4. The fourth estimator gives

$$c_{k4} = \frac{N_h - n_h}{N_h - 1} \frac{n}{n - H} \frac{N - H}{N}.$$

Next,

$$\hat{y}_k^* = \frac{N_h}{n_h} \frac{1}{n_h} \sum_{k \in U_h} y_k S_k = \frac{N_h}{n_h} \widehat{\overline{Y}}_h,$$

and

$$\widehat{\mathrm{var}}_{\mathrm{APPROX4}} \left(\widehat{Y}_{HT} \right) = \frac{n}{n - H} \frac{N - H}{N} \sum_{h=1}^{H} N_h^2 \frac{N_h - n_h}{N_h - 1} \frac{n_h - 1}{n_h} \frac{v_{yh}^2}{n_h}.$$

5. The fifth estimator gives

$$c_{k5} = \frac{n_h}{n_h - 1} \frac{N_h - n_h}{N_h}.$$

Next,

$$\hat{y}_k^* = \frac{N_h}{n_h} \frac{1}{n_h} \sum_{k \in U_h} y_k S_k = \frac{N_h}{n_h} \widehat{\overline{Y}}_h,$$

and

$$\widehat{\mathrm{var}}_{\mathrm{APPROX5}} \left(\widehat{Y}_{HT} \right) = \sum_{h=1}^{H} N_h^2 \frac{N_h - n_h}{N_h} \frac{v_{yh}^2}{n_h}.$$

The five estimators are very similar, but $\widehat{\mathrm{var}}_{\mathrm{APPROX5}}$ is the only approximation that gives the exact variance of a stratified sampling design.

8.10 Recent Developments in Balanced Sampling

Balanced sampling is at present a common procedure of sampling which is used, for instance, for the selection of the master sample at the Institut National de la Statistique et des Études Économiques (INSEE, France). The implementation of the cube method has posed new, interesting problems. The question of the determination of the optimal inclusion probabilities is developed under balanced sampling in Tillé and Favre (2005) and generalizes the optimal stratification. The problem of coordinating several balanced sampling designs is intricate, but several solutions have been given in Tillé and Favre (2004). The delicate question of the use of calibration or balanced sampling is discussed in Deville and Tillé (2004) and in a restricted context in Berger et al. (2003). Deville (2005) has proposed to use the cube method in order to provide balanced imputations for nonresponse.

An Example of the Cube Method

9.1 Implementation of the Cube Method

The cube method was first implemented in SAS-IML® by three students from the École Nationale de la Statistique et de l'Analyse de l'Information (Rennes, France) (see Bousabaa et al., 1999). This prototype was circulated in the Institut National de la Statistique et des Études Économiques (INSEE, France) and was modified by Jean Dumais, Philippe Bertrand, Jean-Michel Grignon, and Fréderic Tardieu (see Tardieu, 2001). Next, the macro was completely rewritten by Pascal Ardilly, Sylvie Rousseau, Guillaume Chauvet, Bernard Weytens Frederic Tardieu and is now available on the INSEE website (see Rousseau and Tardieu, 2004). In the R language, a package has also been developed by Tillé and Matei (2005) which allows selecting unequal probabilities and balanced samples.

This macro allows the selection of samples with unequal probabilities of up to 50'000 units and 30 balancing variables. INSEE has adopted the cube method for its most important statistical projects. For instance, in the redesigned census in France, a fifth of the municipalities with fewer than 10,000 inhabitants are sampled each year, so that after five years all municipalities will be selected. All households in these municipalities are surveyed. The 5 samples of municipalities are selected with equal probabilities using the cube method and are balanced on a set of demographic variables (Dumais and Isnard, 2000). Wilms (2000) has also used the cube method for the selection of the French master sample of households.

A first version of the fast Algorithm 8.4, presented on page 162, has also been implemented in a SAS-IML® macro by Guillaume Chauvet (see Chauvet and Tillé, 2006, 2005a,b). With this faster method, a sample of size 62,740 has been selected in a population of 313,702 units, corresponding to the addresses of the largest municipalities (10,000 inhabitants or more) of the Rhone-Alpes French region. Because the computation time of this new version increases linearly with the population size, the population size is no longer limited.

9.2 The Data of Municipalities in the Canton of Ticino (Switzerland)

The data of municipalities in Ticino were kindly furnished by the Swiss federal statistical office and come from the federal census of population whose date of reference is May 5, 2000. Ticino is made up of 245 municipalities. The list of variables is presented in Table 9.1. The data are presented in the appendix on page 181.

Table 9.1. List of variables for the population of municipalities in Ticino

POP	Number of men and women
ONE	Constant variable that always takes the value 1
ARE	Area of the municipality in hectares
POM	Number of men
POW	Number of women
P00	Number of men and women aged between 0 and 20
P20	Number of men and women aged between 20 and 40
P40	Number of men and women aged between 40 and 65
P65	Number of men and women aged 65 and over
HOU	Number of households

At the time of the census, there is a total of 306,846 inhabitants in the canton of Ticino. The largest municipality is Lugano and has 26 560 inhabitants, and the smallest one is Corippo with 22 inhabitants. The sizes of the municipalities are thus very heterogeneous.

9.3 The Sample of Municipalities

The municipalities are selected with unequal inclusion probabilities proportional to the variable "number of men and women" in the municipalities (variable POP). The sample size is 50. Algorithm 2.1, page 19, has been used to compute the inclusion probabilities. The largest 12 cities are Lugano, Bellinzona, Locarno, Chiasso, Pregassona, Giubiasco, Minusio, Losone, Viganello, Biasca, Mendrisio and Massagno and have an inclusion probability equal to 1. They are thus always selected in the sample. We have used the SAS-IML® macro CUBE that is available on the INSEE website. The selected option for the landing phase is the resolution of the linear program, and the cost is the sum of the coefficients of variation. The selected units are presented in Table 9.2 in decreasing order of population size. The last column (PI) contains the inclusion probabilities.

Table 9.2: Selected sample of municipalities

NUM	MUNI	POP	ARE	POM	POW	P00	P20	P40	P65	HOU	PI
5192	Lugano	26560	1170	11953	14607	4337	7846	8554	5823	13420	1.000
5002	Bellinzona	16463	1915	7701	8762	3170	4709	5432	3152	7294	1.000
5113	Locarno	14561	1929	6678	7883	2929	4166	4730	2736	6730	1.000
5250	Chiasso	7720	533	3584	4136	1257	2186	2538	1739	3774	1.000
5005	Giubiasco	7418	623	3545	3873	1526	2147	2455	1290	3273	1.000
5215	Pregassona	7354	223	3549	3805	1602	2417	2431	904	3211	1.000
5118	Minusio	6428	585	2940	3488	1126	1623	2278	1401	3080	1.000
5234	Viganello	6284	118	2887	3397	1118	1884	2089	1193	3026	1.000
5254	Mendrisio	6146	673	2847	3299	1012	1727	2069	1338	2794	1.000
5115	Losone	5907	953	2900	3007	1194	1828	2068	817	2545	1.000
5281	Biasca	5795	5913	2886	2909	1253	1630	1984	928	2369	1.000
5196	Massagno	5558	73	2552	3006	993	1682	1834	1049	2685	1.000
5091	Ascona	4984	497	2242	2742	853	1154	1834	1143	2472	0.993
5158	Breganzona	4782	227	2250	2532	1058	1333	1607	784	2079	0.953
5257	Morbio Inferiore	4105	229	1934	2171	846	1169	1399	691	1661	0.818
5108	Gordola	3878	703	1886	1992	816	1060	1348	654	1652	0.773
5001	Arbedo-Castione	3729	2128	1862	1867	827	1028	1380	494	1452	0.743
5210	Paradiso	3694	89	1750	1944	625	1202	1235	632	1762	0.736
5141	Agno	3655	249	1710	1945	702	1143	1276	534	1588	0.729
5242	Balerna	3415	257	1647	1768	651	956	1104	704	1519	0.681
5227	Torricella-Taverne	2704	524	1314	1390	684	789	936	295	1036	0.539
5120	Muralto	2676	60	1225	1451	403	667	899	707	1448	0.533
5251	Coldrerio	2538	246	1221	1317	519	717	891	411	1074	0.506
5131	Tenero-Contra	2295	369	1121	1174	481	655	746	413	992	0.457
5282	Claro	2159	2122	1020	1139	527	641	687	304	797	0.430
5017	Sant'Antonino	2066	658	1045	1021	491	669	693	213	794	0.412
5221	Savosa	2061	74	967	1094	346	603	715	397	862	0.411
5224	Sonvico	1600	1106	774	826	325	444	550	281	646	0.319
5189	Lamone	1564	186	767	797	375	496	521	172	632	0.312
5072	Faido	1548	372	725	823	316	365	501	366	614	0.309
5212	Ponte Capriasca	1478	620	714	764	344	436	534	164	594	0.295
5226	Tesserete	1424	309	639	785	307	363	467	287	589	0.284
5217	Rivera	1415	1335	696	719	319	431	479	186	581	0.282
5214	Porza	1348	158	657	691	286	386	506	170	586	0.269
5180	Cureglia	1219	106	601	618	292	350	454	123	497	0.243
5220	Sala Capriasca	1179	842	620	559	268	337	424	150	473	0.235
5010	Lumino	1127	995	556	571	229	307	411	180	470	0.225
5194	Manno	1045	238	536	509	211	334	379	121	414	0.208
5111	Intragna	915	2405	416	499	173	214	309	219	395	0.182
5247	Capolago	758	178	370	388	151	241	238	128	299	0.151
5205	Muzzano	736	157	354	382	154	176	301	105	322	0.147
5008	Gudo	679	994	328	351	133	207	224	115	304	0.135
5161	Cademario	596	396	276	320	114	160	196	126	266	0.119
5181	Curio	521	287	263	258	128	137	167	89	211	0.104
5006	Gnosca	514	745	255	259	99	157	180	78	202	0.102

Table 9.2: (continued)

NUM	MUNI	POP	ARE	POM	POW	P00	P20	P40	P65	HOU	PI
5031	Aquila	487	6296	246	241	108	110	153	116	214	0.097
5123	Piazzogna	362	387	180	182	76	117	114	55	160	0.072
5074	Mairengo	272	665	139	133	66	76	90	40	94	0.054
5038	Leontica	267	738	138	129	46	62	84	75	125	0.053
5122	Palagnedra	92	1674	40	52	7	18	39	28	54	0.018

9.4 Quality of Balancing

The quality of balancing is analyzed in Table 9.3, which contains:

- The population totals for each variable X_j,
- The estimated total by the Horvitz-Thompson estimator $\widehat{X}_{jHT}$,
- The relative deviation in % defined by

$$\text{RD} = 100 \times \frac{\widehat{X}_{jHT} - X_j}{X_j}.$$

Table 9.3. Quality of balancing

Variable	Population Total	Estimated Total	Relative Deviation in %
POP	306846	306846.0	0.00
ONE	245	248.6	1.49
HA	273758	276603.1	1.04
POM	146216	146218.9	0.00
POW	160630	160627.1	-0.00
P00	60886	60653.1	-0.38
P20	86908	87075.3	0.19
P40	104292	104084.9	-0.20
P65	54760	55032.6	0.50
HOU	134916	135396.6	0.36

In spite of the small sample size ($n(\mathbf{S}) = 50$), the quality of balancing is very good. The rounding problem is less than 1% for almost all the variables. This sample will provide very accurate estimators for all the variables that are correlated with the balancing variables.

Appendix: Population of Municipalities in the Canton of Ticino (Switzerland)

Table 9.4: Municipalities in the canton of Ticino, population by gender and age, area, and number of households. Source: Swiss federal statistical office (the data are described in Section 9.2, page 178)

NUM	MUNI	POP	ARE	POM	POW	P00	P20	P40	P65	HOU
5001	Arbedo-Castione	3729	2128	1862	1867	827	1028	1380	494	1452
5002	Bellinzona	16463	1915	7701	8762	3170	4709	5432	3152	7294
5003	Cadenazzo	1755	565	906	849	399	575	581	200	688
5004	Camorino	2210	828	1096	1114	496	623	757	334	934
5005	Giubiasco	7418	623	3545	3873	1526	2147	2455	1290	3273
5006	Gnosca	514	745	255	259	99	157	180	78	202
5007	Gorduno	621	922	294	327	133	196	199	93	258
5008	Gudo	679	994	328	351	133	207	224	115	304
5009	Isone	353	1282	193	160	60	94	134	65	147
5010	Lumino	1127	995	556	571	229	307	411	180	470
5011	Medeglia	330	627	164	166	51	98	111	70	144
5012	Moleno	105	749	50	55	23	30	31	21	41
5013	Monte Carasso	2133	964	1048	1085	460	685	690	298	932
5014	Pianezzo	489	802	231	258	103	122	174	90	220
5015	Preonzo	484	1644	237	247	117	137	173	57	195
5016	Robasacco	108	273	53	55	20	27	42	19	53
5017	Sant'Antonino	2066	658	1045	1021	491	669	693	213	794
5018	Sant'Antonio	168	3358	78	90	31	39	58	40	88
5019	Sementina	2646	825	1290	1356	554	792	928	372	1072
5031	Aquila	487	6296	246	241	108	110	153	116	214
5032	Campo (Blenio)	68	2196	39	29	10	17	25	16	32
5033	Castro	81	307	35	46	20	25	14	22	35
5034	Corzoneso	506	1152	219	287	107	119	130	150	157
5035	Dongio	423	1286	207	216	85	117	130	91	187
5036	Ghirone	44	3027	22	22	7	9	18	10	21
5037	Largario	25	126	14	11	10	4	7	4	8
5038	Leontica	267	738	138	129	46	62	84	75	125

Table 9.4: (Continued)

NUM	MUNI	POP	ARE	POM	POW	P00	P20	P40	P65	HOU
5039	Lottigna	79	651	39	40	13	21	25	20	40
5040	Ludiano	291	619	150	141	54	87	76	74	129
5041	Malvaglia	1172	8030	590	582	229	329	366	248	503
5042	Marolta	43	281	23	20	7	7	17	12	22
5043	Olivone	845	7609	416	429	163	231	256	195	364
5044	Ponto Valentino	218	1029	103	115	33	42	72	71	99
5045	Prugiasco	136	595	71	65	18	33	50	35	57
5046	Semione	320	1042	166	154	64	69	125	62	151
5047	Torre	282	1073	140	142	68	57	88	69	117
5061	Airolo	1593	9437	814	779	288	410	593	302	700
5062	Anzonico	98	1061	46	52	8	20	36	34	48
5063	Bedretto	72	7523	35	37	4	16	16	36	37
5064	Bodio	1058	644	538	520	180	288	374	216	496
5065	Calonico	42	318	17	25	6	11	12	13	21
5066	Calpiogna	40	330	22	18	6	7	16	11	18
5067	Campello	45	396	25	20	8	7	18	12	22
5068	Cavagnago	83	668	44	39	9	10	32	32	44
5069	Chiggiogna	378	392	185	193	87	95	120	76	166
5070	Chironico	403	5777	195	208	91	88	137	87	185
5071	Dalpe	158	1452	80	78	29	25	73	31	70
5072	Faido	1548	372	725	823	316	365	501	366	614
5073	Giornico	885	1948	440	445	162	243	310	170	380
5074	Mairengo	272	665	139	133	66	76	90	40	94
5075	Osco	168	1190	128	40	9	47	78	34	43
5076	Personico	353	3904	175	178	79	82	121	71	148
5077	Pollegio	723	589	393	330	149	207	249	118	284
5078	Prato (Leventina)	397	1685	205	192	81	108	137	71	166
5079	Quinto	1057	7520	523	534	214	272	357	214	433
5080	Rossura	55	1458	31	24	6	8	23	18	28
5081	Sobrio	74	636	37	37	14	10	23	27	38
5091	Ascona	4984	497	2242	2742	853	1154	1834	1143	2472
5092	Auressio	71	299	33	38	19	10	26	16	31
5093	Berzona	48	505	22	26	9	17	12	10	21
5094	Borgnone	143	1060	69	74	24	21	56	42	63
5095	Brione (Verzasca)	203	4856	102	101	41	38	77	47	84
5096	Brione sopra Minusio	484	384	225	259	71	110	200	103	241
5097	Brissago	1833	1779	854	979	284	436	621	492	862
5098	Caviano	111	320	50	61	24	15	46	26	46
5099	Cavigliano	646	548	306	340	147	168	227	104	269
5101	Contone	703	233	343	360	147	246	224	86	307
5102	Corippo	22	773	11	11	1	5	8	8	11
5104	Cugnasco	1120	1703	542	578	273	330	361	156	436
5105	Frasco	100	2573	53	47	25	29	29	17	37
5106	Gerra (Gambarogno)	254	314	124	130	32	59	85	78	127
5107	Gerra (Verzasca)	1098	1868	517	581	225	281	419	173	467
5108	Gordola	3878	703	1886	1992	816	1060	1348	654	1652

Table 9.4: (Continued)

NUM	MUNI	POP	ARE	POM	POW	P00	P20	P40	P65	HOU
5109	Gresso	35	1108	14	21	1	6	13	15	19
5110	Indemini	39	1132	20	19	11	3	15	10	21
5111	Intragna	915	2405	416	499	173	214	309	219	395
5112	Lavertezzo	1098	5811	545	553	263	361	326	148	450
5113	Locarno	14561	1929	6678	7883	2929	4166	4730	2736	6730
5114	Loco	254	907	116	138	32	55	91	76	107
5115	Losone	5907	953	2900	3007	1194	1828	2068	817	2545
5116	Magadino	1499	723	722	777	308	435	492	264	677
5117	Mergoscia	181	1214	89	92	35	51	55	40	84
5118	Minusio	6428	585	2940	3488	1126	1623	2278	1401	3080
5119	Mosogno	57	861	33	24	7	13	25	12	26
5120	Muralto	2676	60	1225	1451	403	667	899	707	1448
5121	Orselina	866	195	354	512	135	154	299	278	399
5122	Palagnedra	92	1674	40	52	7	18	39	28	54
5123	Piazzogna	362	387	180	182	76	117	114	55	160
5125	Ronco sopra Ascona	659	502	322	337	102	130	270	157	343
5127	San Nazzaro	641	553	291	350	104	138	214	185	278
5128	Sant'Abbondio	123	320	53	70	16	28	46	33	59
5129	Sonogno	86	3752	48	38	7	22	38	19	38
5130	Tegna	661	289	332	329	144	180	233	104	274
5131	Tenero-Contra	2295	369	1121	1174	481	655	746	413	992
5132	Vergeletto	65	4078	29	36	3	6	24	32	35
5133	Verscio	887	300	440	447	190	220	349	128	375
5134	Vira (Gambarogno)	616	1199	290	326	97	145	215	159	277
5135	Vogorno	304	2388	151	153	59	69	112	64	147
5136	Onsernone	322	2986	160	162	45	63	114	100	147
5141	Agno	3655	249	1710	1945	702	1143	1276	534	1588
5142	Agra	401	128	199	202	96	105	147	53	159
5143	Aranno	267	258	131	136	58	85	81	43	115
5144	Arogno	969	850	478	491	220	241	300	208	411
5145	Arosio	422	656	215	207	71	140	144	67	188
5146	Astano	290	378	155	135	50	57	107	76	137
5147	Barbengo	1559	266	744	815	404	495	497	163	613
5148	Bedano	1196	187	564	632	245	357	401	193	441
5149	Bedigliora	540	248	261	279	140	137	173	90	233
5150	Bidogno	296	349	142	154	42	86	120	48	138
5151	Bioggio	1504	305	689	815	326	460	499	219	635
5153	Bironico	512	418	258	254	118	162	167	65	212
5154	Bissone	711	181	344	367	112	202	253	144	348
5155	Bogno	93	421	44	49	16	21	29	27	41
5156	Bosco Luganese	348	153	174	174	82	84	129	53	139
5158	Breganzona	4782	227	2250	2532	1058	1333	1607	784	2079
5159	Breno	255	575	122	133	56	49	88	62	126
5160	Brusino Arsizio	454	404	219	235	97	123	151	83	205
5161	Cademario	596	396	276	320	114	160	196	126	266
5162	Cadempino	1317	76	650	667	304	434	442	137	528

Table 9.4: (Continued)

NUM	MUNI	POP	ARE	POM	POW	P00	P20	P40	P65	HOU
5163	Cadro	1797	452	938	859	381	530	676	210	639
5164	Cagiallo	538	551	265	273	124	138	197	79	221
5165	Camignolo	596	453	299	297	124	172	218	82	237
5167	Canobbio	1825	130	885	940	336	601	627	261	806
5168	Carabbia	512	106	248	264	121	125	190	76	195
5169	Carabietta	100	46	51	49	21	23	27	29	44
5170	Carona	681	475	327	354	154	160	264	103	298
5171	Caslano	3495	284	1633	1862	799	943	1208	545	1497
5173	Certara	65	273	31	34	9	17	20	19	31
5174	Cimadera	100	528	48	52	16	22	36	26	55
5175	Cimo	209	86	102	107	39	64	77	29	87
5176	Comano	1594	206	740	854	334	380	614	266	662
5177	Corticiasca	138	215	75	63	33	36	52	17	58
5178	Croglio	865	438	396	469	153	237	297	178	372
5179	Cureggia	112	69	56	56	27	27	44	14	45
5180	Cureglia	1219	106	601	618	292	350	454	123	497
5181	Curio	521	287	263	258	128	137	167	89	211
5182	Davesco-Soragno	1288	249	616	672	303	393	436	156	536
5183	Fescoggia	88	245	41	47	15	15	35	23	40
5184	Gandria	207	343	105	102	28	60	70	49	108
5185	Gentilino	1328	122	644	684	254	345	512	217	576
5186	Grancia	366	63	177	189	95	118	120	33	145
5187	Gravesano	1022	69	508	514	227	297	383	115	402
5188	Iseo	69	100	30	39	13	17	28	11	29
5189	Lamone	1564	186	767	797	375	496	521	172	632
5190	Lopagno	496	537	254	242	95	140	193	68	182
5191	Lugaggia	697	347	353	344	152	213	230	102	286
5192	Lugano	26560	1170	11953	14607	4337	7846	8554	5823	13420
5193	Magliaso	1359	109	619	740	274	360	473	252	553
5194	Manno	1045	238	536	509	211	334	379	121	414
5195	Maroggia	562	100	288	274	92	163	185	122	268
5196	Massagno	5558	73	2552	3006	993	1682	1834	1049	2685
5197	Melano	1102	464	542	560	222	345	373	162	500
5198	Melide	1501	168	695	806	280	436	519	266	740
5199	Mezzovico-Vira	938	1021	466	472	186	279	311	162	367
5200	Miglieglia	215	513	107	108	40	61	79	35	100
5201	Montagnola	2092	314	981	1111	594	478	696	324	810
5202	Monteggio	784	336	388	396	152	214	265	153	352
5203	Morcote	754	280	345	409	103	202	262	187	344
5204	Mugena	141	363	61	80	41	37	39	24	57
5205	Muzzano	736	157	354	382	154	176	301	105	322
5206	Neggio	352	91	173	179	73	110	121	48	137
5207	Novaggio	716	431	319	397	172	174	238	132	315
5208	Origlio	1158	207	560	598	294	297	419	148	462
5209	Pambio-Noranco	570	57	296	274	118	176	210	66	232
5210	Paradiso	3694	89	1750	1944	625	1202	1235	632	1762

Table 9.4: (Continued)

NUM	MUNI	POP	ARE	POM	POW	P00	P20	P40	P65	HOU
5211	Pazzallo	1162	162	556	606	248	392	390	132	511
5212	Ponte Capriasca	1478	620	714	764	344	436	534	164	594
5213	Ponte Tresa	769	41	353	416	119	212	254	184	373
5214	Porza	1348	158	657	691	286	386	506	170	586
5215	Pregassona	7354	223	3549	3805	1602	2417	2431	904	3211
5216	Pura	1040	304	481	559	217	249	379	195	438
5217	Rivera	1415	1335	696	719	319	431	479	186	581
5218	Roveredo (TI)	126	69	55	71	30	32	43	21	49
5219	Rovio	673	553	327	346	135	205	246	87	281
5220	Sala Capriasca	1179	842	620	559	268	337	424	150	473
5221	Savosa	2061	74	967	1094	346	603	715	397	862
5222	Sessa	604	287	291	313	128	127	223	126	265
5223	Sigirino	390	871	210	180	83	129	127	51	166
5224	Sonvico	1600	1106	774	826	325	444	550	281	646
5225	Sorengo	1557	85	715	842	311	427	554	265	618
5226	Tesserete	1424	309	639	785	307	363	467	287	589
5227	Torricella-Taverne	2704	524	1314	1390	684	789	936	295	1036
5228	Vaglio	496	319	243	253	96	153	178	69	197
5229	Valcolla	558	1134	248	310	94	140	181	143	245
5230	Vernate	363	151	169	194	76	104	128	55	172
5231	Vezia	1575	139	768	807	316	479	570	210	671
5232	Vezio	208	367	98	110	46	63	66	33	90
5233	Vico Morcote	250	191	122	128	49	51	113	37	129
5234	Viganello	6284	118	2887	3397	1118	1884	2089	1193	3026
5235	Villa Luganese	467	220	225	242	93	135	169	70	184
5241	Arzo	1010	279	478	532	236	280	340	154	409
5242	Balerna	3415	257	1647	1768	651	956	1104	704	1519
5243	Besazio	501	88	241	260	105	113	203	80	209
5244	Bruzella	183	344	94	89	40	49	60	34	74
5245	Cabbio	173	569	88	85	32	50	52	39	75
5246	Caneggio	343	390	170	173	75	88	126	54	147
5247	Capolago	758	178	370	388	151	241	238	128	299
5248	Casima	61	99	28	33	13	16	24	8	29
5249	Castel San Pietro	1728	805	810	918	345	446	610	327	675
5250	Chiasso	7720	533	3584	4136	1257	2186	2538	1739	3774
5251	Coldrerio	2538	246	1221	1317	519	717	891	411	1074
5252	Genestrerio	827	149	400	427	146	283	278	120	360
5253	Ligornetto	1408	202	672	736	306	422	481	199	571
5254	Mendrisio	6146	673	2847	3299	1012	1727	2069	1338	2794
5255	Meride	293	746	142	151	56	65	109	63	137
5256	Monte	92	241	47	45	26	27	28	11	38
5257	Morbio Inferiore	4105	229	1934	2171	846	1169	1399	691	1661
5258	Morbio Superiore	700	275	337	363	142	205	254	99	270
5259	Muggio	206	839	102	104	27	48	78	53	101
5260	Novazzano	2369	518	1153	1216	482	677	789	421	941
5262	Rancate	1353	231	657	696	266	349	480	258	558

Table 9.4: (Continued)

NUM	MUNI	POP	ARE	POM	POW	P00	P20	P40	P65	HOU
5263	Riva San Vitale	2292	597	1137	1155	511	702	763	316	924
5264	Sagno	238	168	125	113	38	61	94	45	108
5265	Salorino	487	498	242	245	102	106	195	84	198
5266	Stabio	3627	615	1782	1845	834	1136	1186	471	1447
5267	Tremona	393	158	192	201	95	100	137	61	158
5268	Vacallo	2758	162	1344	1414	517	770	989	482	1169
5281	Biasca	5795	5913	2886	2909	1253	1630	1984	928	2369
5282	Claro	2159	2122	1020	1139	527	641	687	304	797
5283	Cresciano	587	1723	298	289	117	196	192	82	255
5284	Iragna	491	1839	243	248	102	157	160	72	190
5285	Lodrino	1461	3150	752	709	315	441	517	188	549
5286	Osogna	941	1899	495	446	224	291	311	115	352
5301	Aurigeno	372	1094	189	183	84	96	142	50	161
5302	Avegno	493	811	243	250	118	131	172	72	191
5303	Bignasco	306	8151	168	138	86	76	104	40	114
5304	Bosco/Gurin	71	2204	39	32	12	22	27	10	35
5305	Broglio	88	1297	41	47	29	26	19	14	34
5306	Brontallo	50	1044	24	26	15	10	18	7	19
5307	Campo (Vallemaggia)	58	4327	28	30	2	7	24	25	30
5308	Cavergno	468	5520	227	241	118	116	141	93	203
5309	Cerentino	58	2007	30	28	12	16	13	17	28
5310	Cevio	497	1477	241	256	93	107	165	132	201
5311	Coglio	96	957	48	48	23	22	37	14	35
5312	Fusio	45	6080	24	21	6	8	15	16	21
5313	Giumaglio	202	1316	96	106	41	53	75	33	81
5314	Gordevio	798	1924	347	451	175	194	261	168	272
5315	Linescio	32	661	14	18	2	3	14	13	19
5316	Lodano	171	1375	87	84	39	40	65	27	68
5317	Maggia	850	2385	398	452	195	203	259	193	314
5318	Menzonio	73	1064	31	42	13	22	27	11	34
5319	Moghegno	336	703	164	172	82	88	117	49	141
5320	Peccia	171	5422	82	89	39	33	59	40	64
5321	Prato-Sornico	104	3837	59	45	21	24	43	16	40
5322	Someo	254	3279	122	132	38	54	89	73	111

List of Tables

List of Figures

List of Algorithms

List of abbreviations

b est var	estimator of variance generally biased
BERN	Bernoulli sampling design
BERNWR	Bernoulli sampling design with replacement
CPS	conditional Poisson sampling
d by d	draw by draw
EP	equal probability
exact	satisfies conjointly the three properties: strpps, strwor, n fixed
exponential	procedure that implements an exponential design
EXT	exponential (design)
FS	fixed size
HH	Hansen-Hurwitz
HT	Horvitz-Thompson
inexact	fails to satisfy at least one of the three properties: strpps, strwor, n fixed
j p enum	calculation of joint inclusion probability in sample involves enumeration of all possible selections, or at least a large number of them
j p iter	calculation of joint inclusion probability in sample involves iteration on computer
LPD	largest possible deviation (of variance)
MULTI	multinomial design
n=2 only	applicable for sample size equal to 2 only
n fixed	number of units in sample fixed
nonrotg	nonrotating
not gen app	not generally applicable
ord	ordered
POISS	Poisson sampling design
POISSWOR	Poisson sampling design without replacement
POISSWR	Poisson sampling design with replacement
rej	rejective

SRS	simple random sampling
SRSWOR	simple random sampling without replacement
SRSWR	simple random sampling with replacement
strpps	inclusion probability strictly proportional to size
strwor	strictly without replacement
syst	systematic
unord	unordered
UP	unequal probability
WOR	without replacement
WR	with replacement
ws	whole sample

Table of Notations

$'$	vector $\mathbf{u}'$ is the transpose of vector $\mathbf{u}$
$!$	factorial : $n! = n \times (n-1) \times \cdots \times 2 \times 1$
$\circ$	$\overset{\circ}{\mathcal{Q}}$ is the interior of support $\mathcal{Q}$
$\binom{N}{n}$	$\frac{N!}{n!(N-n)!}$ number of combinations of n objects from a set of N objects
$\approx$	is approximately equal
$\sim$	follows a specified distribution (for a random variable)
$\overrightarrow{\mathcal{Q}}$	direction of a support $\mathcal{Q}$
$\mathbf{0}$	null vector $(0 \cdots 0 \cdots 0)'$ of $\mathbb{R}^N$
$\mathbf{1}$	vector $(1 \cdots 1 \cdots 1)'$ of $\mathbb{R}^N$
$\mathbf{A}$	$p \times N$ matrix of the weighted vectors $(\check{\mathbf{x}}_1 \cdots \check{\mathbf{x}}_N)$
$\mathrm{B}(\widehat{T})$	bias of estimator $\widehat{T}$
$\mathcal{B}(n, p)$	binomial random variable with parameters n and p
card	cardinal (number of units of a set)
$\mathrm{cov}(X, Y)$	covariance between the random variables X and Y
d	possible value for the observed data
$\mathbb{C}$	set of the complex numbers
D	observed data
$\mathrm{E}(Y)$	expected value of random variable Y
$\mathrm{E}(Y\vert A)$	conditional expected value of random variable Y given A
$\mathrm{EXT}(K)$	set of the external points of a polytope K
$\mathbf{H}$	projection matrix $\mathbf{H} = \mathbf{I} - \mathbf{1}\mathbf{1}'/N$ that centers the data
$\mathbf{I}$	identity matrix of dimension $N \times N$
Invariant$\mathcal{Q}$	invariant subspace spanned by a support $\mathcal{Q}$
k	usually denotes the label of a unit of the population
$\mathrm{Ker}\mathbf{A}$	kernel of matrix $\mathbf{A}$; that is, $\{\mathbf{u}\vert\mathbf{A}\mathbf{u} = \mathbf{0}\}$

LPD	least possible deviation between the variance of two sampling designs
MSE	mean square error
$n(\mathbf{s})$	size of sample $\mathbf{s}$
N	size of the population
$\mathbb{N}$	set of all natural numbers $\mathbb{N} = \{0, 1, 2, 3, 4, \dots\}$
$\mathbb{N}^*$	set of the positive natural numbers $\mathbb{N}^* = \{1, 2, 3, 4, \dots\}$
$N(\mu, \sigma^2)$	normal random variable with mean μ and variance σ^2
p	number of auxiliary or balancing variables
$p(\mathbf{s})$	probability of selecting sample $\mathbf{s}$
$\Pr(A)$	probability of event A
$\Pr(A|B)$	conditional probability of event A given B
$\mathcal{P}(\lambda)$	Poisson random variable with parameter λ
$\mathcal{Q}$	support; that is, set of samples
$\mathbb{R}$	set of the real numbers
$\mathbb{R}_+$	set of the nonnegative real numbers $\mathbb{R}_+ = \{x \in \mathbb{R} | x \geq 0\}$
$\mathbb{R}_+^*$	set of the positive real numbers $\mathbb{R}_+ = \{x \in \mathbb{R} | x > 0\}$
$\mathcal{R}$	support of all the samples with replacement $\mathcal{R} = \mathbb{N}^N$
$\mathcal{R}_n$	support of all the samples with replacement of fixed size n
$\mathbf{s}$	sample, $\mathbf{s} \subset \mathbb{N}^N$
s_k	number of times that unit k appears in the sample $\mathbf{s}$
$\mathbf{S}$	random sample such that $\Pr(\mathbf{S} = \mathbf{s}) = p(\mathbf{s})$
S_k	number of times that unit k appears in the random sample $\mathbf{S}$
$\mathcal{S}$	support of all the samples without replacement $\mathcal{S} = \{0, 1\}^N$
$\mathcal{S}_n$	support of all the samples without replacement of fixed size n
$\mathfrak{s}_m^{(z)}$	Stirling number of the second kind
U	set of the labels of the units in the population: $U = \{1, \dots, k, \dots, N\}$
$\mathcal{U}[0, 1[$	uniform random variable in $[0,1[$
v_y^2	corrected variance of variable y of the sample
V_y^2	corrected variance of variable y of the population
$\text{var}(Y)$	variance of random variable Y
$\widehat{\text{var}}(Y)$	estimator of variance of the random variable Y
x_j	jth auxiliary or balancing variable
x_{kj}	value taken by variable x_j for unit k
$\mathbf{x}_k$	vector $(x_{k1} \ \cdots \ x_{kj} \ \cdots \ x_{kp})'$
$\check{\mathbf{x}}_k$	weighted vector $\mathbf{x}_k/\pi_k$
y	variable of interest
y_k	value taken by variable y for unit k
$\check{y}_k$	weighted value y_k/π_k
Y	total of the values taken by variable y on U
$\widehat{Y}_{HT}$	Horvitz-Thompson estimator of Y

$\widehat{Y}_{HT}$	Hansen-Hurwitz estimator of Y
$\overline{Y}$	mean of the values taken by variables Y on the units of U
$\Delta_{k\ell}$	$\pi_{k\ell} - \pi_k\pi_\ell$
$\boldsymbol{\Delta}$	matrix of $\Delta_{k\ell}$
θ	parameter of a simple design
λ	parameter of an exponential design
μ_k	expected value of S_k
$\mu_{k\ell}$	expected value of $S_k S_\ell$
$\boldsymbol{\mu}$	vector of expected value of $\mathbf{S}$
$\boldsymbol{\pi}$	vector of inclusion probabilities
π_k	inclusion probability of k
$\pi_{k\ell}$	joint inclusion probability of k and ℓ
$\boldsymbol{\Pi}$	matrix of joint inclusion probabilities
σ_y^2	variance of y on the population
$\boldsymbol{\Sigma}$	variance-covariance matrix of $\mathbf{S}$
$\phi(\mathbf{t})$	characteristic function of a sampling design

References

Abramowitz, M. and Stegun, I.A. (1964). *Handbook of Mathematical Functions*. New York: Dover.

Ahrens, J.H. and Dieter, U. (1985). Sequential random sampling. *ACM Transactions on Mathematical Software*, **11**, 157–169.

Aires, N. (1999). Algorithms to find exact inclusion probabilities for conditional Poisson sampling and Pareto πps sampling designs. *Methodology and Computing in Applied Probability*, **4**, 457–469.

Aires, N. (2000a). Comparisons between conditional Poisson sampling and Pareto πps sampling designs. *Journal of Statistical Planning and Inference*, **82**, 1–15.

Aires, N. (2000b). *Techniques to calculate exact inclusion probabilities for conditional Poisson sampling and Pareto πps sampling designs*. Doctoral thesis, Chalmers University of Technology and Göteborg Univerity, Göteborg, Sweden.

Ardilly, P. (1991). Échantillonnage représentatif optimum à probabilités inégales. *Annales d'Économie et de Statistique*, **23**, 91–113.

Asok, C. and Sukhatme, B.V. (1976). On Sampford's procedure of unequal probability sampling without replacement. *Journal of the American Statistical Association*, **71**, 912–918.

Avadhani, M.S. and Srivastava, A.K. (1972). A comparison of Midzuno-Sen scheme with PPS sampling without replacement and its application to successive sampling. *Annals of the Institute of Statistical Mathematics*, **24**, 153–164.

Basu, D. (1958). On sampling with and without replacement. *Sankhyā*, **20**, 287–294.

Basu, D. (1969). Role of the sufficiency and likelihood principles in sample survey theory. *Sankhyā*, **A31**, 441–454.

Basu, D. and Ghosh, J.K. (1967). Sufficient statistics in sampling from a finite universe. *Pages 850–859 of: Proceedings of the 36th Session of International Statistical Institute*.

Bayless, D.L. and Rao, J.N.K. (1970). An empirical study of stabilities of estimators and variance estimators in unequal probability sampling ($n = 3$ or 4). *Journal of the American Statistical Association*, **65**, 1645–1667.

Bebbington, A.C. (1975). A simple method of drawing a sample without replacement. *Applied Statistics*, **24**, 136.

Bellhouse, D.R. (1988). Systematic Sampling. *Pages 125–145 of:* Krishnaiah, P.R. and Rao, C.R. (eds.), *Handbook of Statistics Volume 6: Sampling*. Amsterdam: Elsevier/North-Holland.

Bellhouse, D.R. and Rao, J.N.K. (1975). Systematic sampling in the presence of a trend. *Biometrika*, **62**, 694–697.

Berger, Y.G. (1998a). Rate of convergence for asymptotic variance for the Horvitz-Thompson estimator. *Journal of Statistical Planning and Inference*, **74**, 149–168.

Berger, Y.G. (1998b). Variance estimation using list sequential scheme for unequal probability sampling. *Journal of Official Statistics*, **14**, 315–323.

Berger, Y.G. (2003). A modified Hájek variance estimator for systematic sampling. *Statistics in Transition*, **6**, 5–21.

Berger, Y.G., El Haj Tirari, M. and Tillé, Y. (2003). Toward optimal regression estimation in sample surveys. *Australian and New-Zealand Journal of Statistics*, **45**, 319–329.

Bethlehem, J.G. and Schuerhoff, M.H. (1984). Second-order inclusion probabilities in sequential sampling without replacement with unequal probabilities. *Biometrika*, **71**, 642–644.

Bissell, A.F. (1986). Ordered random selection without replacement. *Applied Statistics*, **35**, 73–75.

Bol'shev, L.N. (1965). On a characterization of the Poisson distribution. *Teoriya Veroyatnostei i ee Primeneniya*, **10**, 64–71.

Bondesson, L., Traat, I. and Lundqvist, A. (2004). *Pareto sampling versus Sampford and conditional Poisson sampling*. Tech. rept. 6. Umeå University, Sweden.

Bousabaa, A., Lieber, J. and Sirolli, R. (1999). *La Macro Cube*. Tech. rept. INSEE, Rennes.

Brewer, K.R.W. (1963a). A model of systematic sampling with unequal probabilities. *Australian Journal of Statistics*, **5**, 5–13.

Brewer, K.R.W. (1963b). Ratio estimation in finite populations: Some results deductible from the assumption of an underlying stochastic process. *Australian Journal of Statistics*, **5**, 93–105.

Brewer, K.R.W. (1975). A simple procedure for πpswor. *Australian Journal of Statistics*, **17**, 166–172.

Brewer, K.R.W. (1999). Design-based or prediction-based inference? Stratified random vs stratified balanced sampling. *International Statistical Review*, **67**, 35–47.

Brewer, K.R.W. (2001). Deriving and estimating an approximate variance for the Horvitz-Thompson estimator using only first order inclusion probabilities. *In: Proceedings of the Second International Conference on Etablish-ment Surveys. Contributed paper in section 30 on the included CD.*

Brewer, K.R.W. (2002). *Combined Survey Sampling Inference, Weighing Basu's Elephants.* London: Arnold.

Brewer, K.R.W. and Donadio, M.E. (2003). The high entropy variance of the Horvitz-Thompson estimator. *Survey Methodology*, **29**, 189–196.

Brewer, K.R.W. and Hanif, M. (1983). *Sampling with Unequal Probabilities.* New York: Spinger-Verlag.

Brewer, K.R.W, Early, L.J. and Joyce, S.F. (1972). Selecting several samples from a single population. *Australian Journal of Statistics*, **3**, 231–239.

Brewer, K.R.W, Early, L.J. and Hanif, M. (1984). Poisson, modified Poisson and collocated sampling. *Journal of Statistical Planning and Inference*, **10**, 15–30.

Brown, L.D. (1986). *Fundamentals of Statistical Exponential Families: With Applications in Statistical Decision Theory.* Hayward, CA: Institute of Mathematical Statistics.

Brown, M.B. and Bromberg, J. (1984). An efficient two-stage procedure for generating random variates from the multinomial distribution. *The American Statistician*, **38**, 216–219.

Carroll, J.L. and Hartley, H.O. (1964). The symmetric method of unequal probability sampling without replacement. *Biometrics*, **20**, 908–909.

Cassel, C.-M., Särndal, C.-E. and Wretman, J.H. (1993). *Foundations of Inference in Survey Sampling.* New York: Wiley.

Chao, M.T. (1982). A general purpose unequal probability sampling plan. *Biometrika*, **69**, 653–656.

Chaudhuri, A. (1974). On some properties of the sampling scheme due to Midzuno. *Calcutta Statistical Association Bulletin*, **23**, 1–19.

Chauvet, G. and Tillé, Y. (2005a). *Fast SAS Macros for balancing Samples: user's guide.* Software Manual, University of Neuchâtel.

Chauvet, G. and Tillé, Y. (2005b). New SAS macros for balanced sampling. *In: Journées de Méthodologie Statistique, INSEE.*

Chauvet, G. and Tillé, Y. (2006). A fast algorithm of balanced sampling. *To appear in Journal of Computational Statistics.*

Chen, S.X. (1998). Weighted polynomial models and weighted sampling schemes for finite population. *Annals of Statistics*, **26**, 1894–1915.

Chen, S.X. (2000). General properties and estimation of conditional Bernoulli models. *Journal of Multivariate Analysis*, **74**, 67–87.

Chen, S.X. and Liu, J.S. (1997). Statistical applications of the Poisson-binomial and conditional Bernoulli distributions. *Statistica Sinica*, **7**, 875–892.

Chen, S.X., Dempster, A.P. and Liu, J.S. (1994). Weighted finite population sampling to maximize entropy. *Biometrika*, **81**, 457–469.

Chikkagoudar, M.S. (1966). A note on inverse sampling with equal probabilities. *Sankhyā*, **A28**, 93–96.

Choudhry, G.H. (1979). Selecting a sample of size n with PPSWOR from a finite population. *Survey Methodology*, **5**, 79–95.

Chromy, J.R. (1979). Sequential sample selection methods. *Pages 401–406 of: Proceedings of the American Statistical Association, Survey Research Methods Section.*

Cochran, W.G. (1946). Relative accuracy of systematic and stratified random samples for a certain class of population. *Annals of Mathematical Statistics*, **17**, 164–177.

Connor, W.S. (1966). An exact formula for the probability that specified sampling units will occur in a sample drawn with unequal probabilities and without replacement. *Journal of the American Statistical Association*, **61**, 384–490.

Dagpunar, J. (1988). *Principles of Random Numbers Generation.* Oxford, England: Clarendon.

Das, M.N. and Mohanty, S. (1973). On PPS sampling without replacement ensuring selection probabilities exactly proportional to sizes. *Australian Journal of Statistics*, **15**, 87–94.

Davis, C.S. (1993). The computer generation of multinomial random variables. *Computational Statistics and Data Analysis*, **16**, 205–217.

Deville, J.-C. (1992). Constrained samples, conditional inference, weighting: Three aspects of the utilisation of auxiliary information. *In: Proceedings of the Workshop on the Uses of Auxiliary Information in Surveys, Örebro (Sweden).*

Deville, J.-C. (1993). *Estimation de la variance pour les enquêtes en deux phases.* Manuscript. INSEE, Paris.

Deville, J.-C. (1998). *Une nouvelle (encore une!) méthode de tirage à probabilités inégales.* Tech. rept. 9804. Méthodologie Statistique, INSEE.

Deville, J.-C. (1999). Variance estimation for complex statistics and estimators: linearization and residual techniques. *Survey Methodology*, **25**, 193–204.

Deville, J.-C. (2000). *Note sur l'algorithme de Chen, Dempster et Liu.* Tech. rept. CREST-ENSAI, Rennes.

Deville, J.-C. (2005). *Imputation stochastique et échantillonnage équilibré.* Tech. rept. École Nationale de la Statistique et de l'Analyse de l'Information.

Deville, J.-C. and Tillé, Y. (1998). Unequal probability sampling without replacement through a splitting method. *Biometrika*, **85**, 89–101.

Deville, J.-C. and Tillé, Y. (2000). Selection of several unequal probability samples from the same population. *Journal of Statistical Planning and Inference*, **86**, 215–227.

Deville, J.-C. and Tillé, Y. (2004). Efficient balanced sampling: The cube method. *Biometrika*, **91**, 893–912.

Deville, J.-C. and Tillé, Y. (2005). Variance approximation under balanced sampling. *Journal of Statistical Planning and Inference*, **128**, 411–425.

Deville, J.-C., Grosbras, J.-M. and Roth, N. (1988). Efficient sampling algorithms and balanced sample. *Pages 255–266 of: COMPSTAT, Proceedings in Computational Statistics*. Heidelberg: Physica Verlag.

Devroye, L. (1986). *Non-uniform Random Variate Generation*. New York: Spinger-Verlag.

Dumais, J. and Isnard, M. (2000). Le sondage de logements dans les grandes communes dans le cadre du recensement rénové de la population. *Pages 37–76 of: Séries INSEE Méthodes: Actes des Journées de Méthodologie Statistique*, vol. 100. Paris: INSEE.

Durbin, J. (1953). Some results in sampling when the units are selected with unequal probabilities. *Journal of the Royal Statistical Society*, **15**, 262–269.

Fan, C.T., Muller, M.E. and Rezucha, I. (1962). Development of sampling plans by using sequential (item by item) selection techniques and digital computer. *Journal of the American Statistical Association*, **57**, 387–402.

Fellegi, I.P. (1963). Sampling with varying probabilities without replacement: Rotation and non-rotating samples. *Journal of the American Statistical Association*, **58**, 183–201.

Gabler, S. (1981). A comparison of Sampford's sampling procedure versus unequal probability sampling with replacement. *Biometrika*, **68**, 725–727.

Goodman, R. and Kish, L. (1950). Controlled selection - A technique in probability sampling. *Journal of the American Statistical Association*, **45**, 350–372.

Grundy, P.M. (1954). A method of sampling with probability exactly proportional to size. *Journal of the Royal Statistical Society*, **16**, 236–238.

Hájek, J. (1964). Asymptotic theory of rejective sampling with varying probabilities from a finite population. *Annals of Mathematical Statistics*, **35**, 1491–1523.

Hájek, J. (1981). *Sampling from a Finite Population*. New York: Marcel Dekker.

Hansen, M.H. and Hurwitz, W.N. (1943). On the theory of sampling from finite populations. *Annals of Mathematical Statistics*, **14**, 333–362.

Hanurav, T.V. (1967). Optimum utilization of auxiliary information: πps sampling of two units from a stratum. *Journal of the Royal Statistical Society*, **B29**, 374–391.

Hartley, H.O. and Chakrabarty, R.P. (1967). Evaluation of approximate variance formulas in sampling with unequal probabilities and without replacement. *Sankhyā*, **B29**, 201–208.

Hartley, H.O. and Rao, J.N.K. (1962). Sampling with unequal probabilities and without replacement. *Annals of Mathematical Statistics*, **33**, 350–374.

Hedayat, A.S. and Majumdar, D. (1995). Generating desirable sampling plans by the technique of trade-off in experimental design. *Journal of Statistical Planning and Inference*, **44**, 237–247.

Hedayat, A.S. and Sinha, B.K. (1991). *Design and Inference Finite Population Sampling*. New York: Wiley.

Hedayat, A.S., Bing-Ying, L. and Stufken, J. (1989). The construction of ΠPS sampling designs through a method of emptying boxes. *Annals of Statistics*, **4**, 1886–1905.

Hidiroglou, M.A. and Gray, G.B. (1980). Construction of joint probability of selection for systematic P.P.S. sampling. *Applied Statistics*, **29**, 107–112.

Ho, F.C.M., Gentle, J.E. and Kennedy, W.J. (1979). Generation of random variables from the multinomial distribution. *Pages 336–339 of: Proceedings of the American Statistical Association, Statistical Computing Section.*

Hodges, J.L.Jr. and LeCam, L. (1960). The Poisson approximation to the Poisson binomial distribution. *Annals of Mathematical Statistics*, **31**, 737–740.

Horvitz, D.G. and Thompson, D.J. (1952). A generalization of sampling without replacement from a finite universe. *Journal of the American Statistical Association*, **47**, 663–685.

Iachan, R. (1982). Systematic sampling: A critical review. *International Statistical Review*, **50**, 293–303.

Iachan, R. (1983). Asymptotic theory of systematic sampling. *Annals of Statistics*, **11**, 959–969.

Isaki, C.T. and Fuller, W.A. (1982). Survey design under a regression population model. *Journal of the American Statistical Association*, **77**, 89–96.

Jessen, R.J. (1969). Some methods of probability non-replacement sampling. *Journal of the American Statistical Association*, **64**, 175–193.

Johnson, N.L., Kotz, S. and Balakrishnan, N. (1997). *Discrete Multivariate Distributions*. New York: Wiley.

Jonasson, J. and Nerman, O. (1996 April). *On maximum entropy πps-sampling with fixed sample size*. Tech. rept. Göteborg University, Sweden.

Kemp, C.D. and Kemp, A.W. (1987). Rapid generation of frequency tables. *Applied Statistics*, **36**, 277–282.

Knuth, D.E. (1981). *The Art of Computer Programming (Volume II): Seminumerical Algorithms*. Reading, MA: Addison-Wesley.

Konijn, H.S. (1973). *Statistical Theory of Sample Survey Design and Analysis*. Amsterdam: North-Holland.

Korwar, R.M. (1996). One-pass selection of a sample with probability proportional to aggregate size. *Indian Journal of Statistics*, **58B**, 80–83.

Kullback, S. (1959). *Information Theory and Applications*. New York: Wiley.

Loukas, S. and Kemp, C.D. (1983). On computer sampling from trivariate and multivariate discrete distribution. *Journal of Statistical Computation and Simulation*, **17**, 113–123.

Madow, L.H. and Madow, W.G. (1944). On the theory of systematic sampling. *Annals of Mathematical Statistics*, **15**, 1–24.

Madow, W.G. (1949). On the theory of systematic sampling, II. *Annals of Mathematical Statistics*, **20**, 333–354.

Matei, A. and Tillé, Y. (2006). Evaluation of variance approximations and estimators in maximum entropy sampling with unequal probability and fixed sample size. *To appear in Journal of Official Statistics*.

McLeod, A.I. and Bellhouse, D.R. (1983). A convenient algorithm for drawing a simple random sampling. *Applied Statistics*, **32**, 182–184.

Midzuno, H. (1950). An outline of the theory of sampling systems. *Annals of the Institute of Statistical Mathematics*, **1**, 149–156.

Mittelhammer, R.C. (1996). *Mathematical Statistics for Economics and Business*. New York: Spinger-Verlag.

Mukhopadhyay, P. (1972). A sampling scheme to realise a pre-assigned set of inclusion probabilities of first two orders. *Calcutta Statistical Association Bulletin*, **21**, 87–122.

Neyman, J. (1934). On the two different aspects of representative method: The method of stratified sampling and the method of purposive selection. *Journal of the Royal Statistical Society*, **97**, 558–606.

Ogus, J.L. and Clark, D.F. (1971). *The annual survey of manufactures: A report on methodology.* Technical paper no. 24. Bureau of the Census, Washington D.C.

Pathak, P.K. (1962). On simple random sampling with replacement. *Sankhyā*, **A24**, 287–302.

Pathak, P.K. (1988). Simple random sampling. *Pages 97–109 of:* Krishnaiah, P.R. and Rao, C.R. (eds.), *Handbook of Statistics Volume 6: Sampling.* Amsterdam: Elsevier/North-Holland.

Péa, J. and Tillé, Y. (2005). *Is systematic sampling a minimum support design?* Tech. rept. University of Neuchâtel.

Pinciaro, S.J. (1978). An algorithm for calculating joint inclusion probabilities under PPS systematic sampling. *Pages 740–740 of: ASA Proceedings of the Section on Survey Research Methods.* American Statistical Association.

Pinkham, R.S. (1987). An efficient algorithm for drawing a simple random sample. *Applied Statistics*, **36**, 370–372.

Qualité, L. (2005). *A comparison of conditional Poisson sampling versus unequal probability sampling with replacement.* Tech. rept. University of Neuchâtel.

Raj, D. (1965). Variance estimation in randomized systematic sampling with probability proportionate to size. *Journal of the American Statistical Association*, **60**, 278–284.

Raj, D. and Khamis, S.D. (1958). Some remarks on sampling with replacement. *Annals of Mathematical Statistics*, **29**, 550–557.

Rao, J.N.K. (1963a). On three procedures of unequal probability sampling without replacement. *Journal of the American Statistical Association*, **58**, 202–215.

Rao, J.N.K. (1963b). On two systems of unequal probability sampling without replacement. *Annals of the Institute of Statistical Mathematics*, **15**, 67–72.

Rao, J.N.K. (1965). On two simple schemas of unequal probability sampling without replacement. *Journal of the Indian Statistical Association*, **3**, 173–180.

Rao, J.N.K. and Bayless, D.L. (1969). An empirical study of the stabilities of estimators and variance estimators in unequal probability sampling of two units per stratum. *Journal of the American Statistical Association*, **64**, 540–549.

Rao, J.N.K. and Singh, M.P. (1973). On the choice of estimator in survey sampling. *Australian Journal of Statistics*, **2**, 95–104.

Richardson, S.C. (1989). One-pass selection of a sample with probability proportional to size. *Applied Statistics*, **38**, 517–520.

Rosén, B. (1972). Asymptotic theory for successive sampling I. *Annals of Mathematical Statistics*, **43**, 373–397.

Rosén, B. (1991). *Variance estimation for systematic pps-sampling*. Tech. rept. 1991:15. Statistics Sweden.

Rosén, B. (1997). On sampling with probability proportional to size. *Journal of Statistical Planning and Inference*, **62**, 159–191.

Rousseau, S. and Tardieu, F. (2004). *La macro SAS CUBE d'échantillonnage équilibré, Documentation de l'utilisateur*. Tech. rept. INSEE, PARIS.

Royall, R.M. and Herson, J. (1973a). Robust estimation in finite populations I. *Journal of the American Statistical Association*, **68**, 880–889.

Royall, R.M. and Herson, J. (1973b). Robust estimation in finite populations II: Stratification on a size variable. *Journal of the American Statistical Association*, **68**, 891–893.

Sadasivan, G. and Sharma, S. (1974). Two dimensional varying probability sampling without replacement. *Sankhyā*, **C36**, 157–166.

Sampford, M.R. (1967). On sampling without replacement with unequal probabilities of selection. *Biometrika*, **54**, 499–513.

Särndal, C.-E., Swensson, B. and Wretman, J.H. (1992). *Model Assisted Survey Sampling*. New York: Spinger-Verlag.

Scott, A.J., Brewer, K.R.W. and Ho, E.W.H. (1978). Finite population sampling and robust estimation. *Journal of the American Statistical Association*, **73**, 359–361.

Sen, A.R. (1953). On the estimate of the variance in sampling with varying probabilities. *Journal of the Indian Society of Agricultural Statistics*, **5**, 119–127.

Sengupta, S. (1989). On Chao's unequal probability sampling plan. *Biometrika*, **76**, 192–196.

Shao, J. (2003). *Mathematical Statistics*. New York: Spinger-Verlag.

Sinha, B.K. (1973). On sampling schemes to realize preassigned sets of inclusion probabilities of first two orders. *Calcutta Statistical Association Bulletin*, **22**, 89–110.

Slanta, G.G. and Fagan, J.T. (1997). *A modified approach to sample selection and variance estimation with probability proportional to size and fixed sample size*. Tech. rept. MCD Working Paper number: Census/MCD/WP-97/02.

Slanta, J.G. (1999). Implementation of the modified Tillé's sampling procedure in the MECS and R&D surveys. *Pages 64–69 of: ASA Proceedings of the Business and Economic Statistics Section*.

Slanta, J.G. and Kusch, G.L. (2001). A comparison of a modified Tillé sampling procedure to Poisson sampling. *In: Proceedings of Statistics Canada Symposium 2001.*

Stehman, V. and Overton, W. S. (1994). Comparison of variance estimators of the Horvitz-Thompson estimator for randomized variable probability systematic sampling. *Journal of the American Statistical Association,* **89**, 30–43.

Sugden, R.A., Smith, T.M.F. and Brown, R.P. (1996). Chao's list sequential scheme for unequal probability sampling. *Journal of Applied Statistics,* **23**, 413–421.

Sunter, A. (1977). List sequential sampling with equal or unequal probabilities without replacement. *Applied Statistics,* **26**, 261–268.

Sunter, A. (1986). Solutions to the problem of unequal probability sampling without replacement. *International Statistical Review,* **54**, 33–50.

Tardieu, F. (2001). *Échantillonnage équilibré: de la théorie à la pratique.* Tech. rept. INSEE, Paris.

Thionet, P. (1953). *La théorie des sondages.* Paris: INSEE, Imprimerie nationale.

Thompson, M.E. and Seber, G.A.F. (1996). *Adaptive Sampling.* New York: Wiley Series in Probability and Statistics.

Tillé, Y. (1996a). An elimination procedure of unequal probability sampling without replacement. *Biometrika,* **83**, 238–241.

Tillé, Y. (1996b). A moving stratification algorithm. *Survey Methodology,* **22**, 85–94.

Tillé, Y. (2001). *Théorie des sondages: échantillonnage et estimation en populations finies.* Paris: Dunod.

Tillé, Y. and Favre, A.-C. (2004). Co-ordination, combination and extension of optimal balanced samples. *Biometrika,* **91**, 913–927.

Tillé, Y. and Favre, A.-C. (2005). *Optimal allocation in balanced sampling.* Tech. rept. University of Neuchâtel.

Tillé, Y. and Matei, A. (2005). *The R package Sampling.* The Comprehensive R Archive Network, http://cran.r-project.org/, Manual of the Contributed Packages.

Traat, I. (1997). *Sampling design as a multivariate distribution.* Research report No 17, Umeå University, Sweden.

Traat, I., Bondesson, L. and Meister, K. (2004). Sampling design and sample selection through distribution theory. *Journal of Statistical Planning and Inference,* **123**, 395–413.

Tschuprow, A. (1923). On the mathematical expectation of the moments of frequency distributions in the case of correlated observation. *Metron,* **3**, 461–493, 646–680.

Valliant, R., Dorfman, A.H. and Royall, R.M. (2000). *Finite Population Sampling and Inference: A Prediction Approach.* New York: Wiley.

Vijayan, K. (1968). An exact πps sampling scheme, generalization of a method of Hanurav. *Journal of the Royal Statistical Society,* **B30**, 556–566.

Vitter, J.S. (1984). Faster methods for random sampling. *Communications of the ACM*, **27**, 703–718.

Vitter, J.S. (1985). Random sampling with a reservoir. *ACM Transactions on Mathematical Software*, **11**(1), 37–57.

Vitter, J.S. (1987). An efficient algorithm for sequential random sampling. *ACM Transactions on Mathematical Software*, **13**, 58–67.

Wilms, L. (2000). Présentation de l'échantillon-maître en 1999 et application au tirage des unités primaires par la macro cube. *In: Séries INSEE Méthodes: Actes des Journées de Méthodologie Statistique*. Paris: INSEE.

Wolter, K.M. (1984). An investigation of some estimators of variance for systematic sampling. *Journal of the American Statistical Association*, **79**, 781–790.

Wynn, H.P. (1977). Convex sets of finite population plans. *Annals of Statistics*, **5**, 414–418.

Yates, F. (1946). A review of recent statistical developments in sampling and sampling surveys. *Journal of the Royal Statistical Society*, **A109**, 12–43.

Yates, F. (1949). *Sampling Methods for Censuses and Surveys*. London: Griffin.

Yates, F. and Grundy, P.M. (1953). Selection without replacement from within strata with probability proportional to size. *Journal of the Royal Statistical Society*, **B15**, 235–261.

Author Index

Index

Springer Series in Statistics   *(continued from p. ii)*

Huet/Bouvier/Poursat/Jolivet: Statistical Tools for Nonlinear Regression: A Practical
  Guide with S-PLUS and R Examples, 2nd edition.
Ibrahim/Chen/Sinha: Bayesian Survival Analysis.
Jolliffe: Principal Component Analysis, 2nd edition.
Knottnerus: Sample Survey Theory: Some Pythagorean Perspectives.
Kolen/Brennan: Test Equating: Methods and Practices.
Kotz/Johnson (Eds.): Breakthroughs in Statistics Volume I.
Kotz/Johnson (Eds.): Breakthroughs in Statistics Volume II.
Kotz/Johnson (Eds.): Breakthroughs in Statistics Volume III.
Küchler/Sørensen: Exponential Families of Stochastic Processes.
Kutoyants: Statistical Influence for Ergodic Diffusion Processes.
Lahiri: Resampling Methods for Dependent Data.
Le Cam: Asymptotic Methods in Statistical Decision Theory.
Le Cam/Yang: Asymptotics in Statistics: Some Basic Concepts, 2nd edition.
Liu: Monte Carlo Strategies in Scientific Computing.
Longford: Models for Uncertainty in Educational Testing.
Manski: Partial Identification of Probability Distributions.
Mielke/Berry: Permutation Methods: A Distance Function Approach.
Molenberghs/Verbeke: Models for Discrete Longitudinal Data.
Nelsen: An Introduction to Copulas. 2nd edition
Pan/Fang: Growth Curve Models and Statistical Diagnostics.
Parzen/Tanabe/Kitagawa: Selected Papers of Hirotugu Akaike.
Politis/Romano/Wolf: Subsampling.
Ramsay/Silverman: Applied Functional Data Analysis: Methods and Case Studies.
Ramsay/Silverman: Functional Data Analysis, 2nd edition.
Rao/Toutenburg: Linear Models: Least Squares and Alternatives.
Reinsel: Elements of Multivariate Time Series Analysis. 2nd edition.
Rosenbaum: Observational Studies, 2nd edition.
Rosenblatt: Gaussian and Non-Gaussian Linear Time Series and Random Fields.
Särndal/Swensson/Wretman: Model Assisted Survey Sampling.
Santner/Williams/Notz: The Design and Analysis of Computer Experiments.
Schervish: Theory of Statistics.
Shao/Tu: The Jackknife and Bootstrap.
Simonoff: Smoothing Methods in Statistics.
Singpurwalla and Wilson: Statistical Methods in Software Engineering: Reliability and
  Risk.
Small: The Statistical Theory of Shape.
Sprott: Statistical Inference in Science.
Stein: Interpolation of Spatial Data: Some Theory for Kriging.
Taniguchi/Kakizawa: Asymptotic Theory of Statistical Inference for Time Series.
Tanner: Tools for Statistical Inference: Methods for the Exploration of Posterior
  Distributions and Likelihood Functions, 3rd edition.
Tillé: Sampling Algorithms.
van der Laan: Unified Methods for Censored Longitudinal Data and Causality.
van der Vaart/Wellner: Weak Convergence and Empirical Processes: With Applications
  to Statistics.
Verbeke/Molenberghs: Linear Mixed Models for Longitudinal Data.
Weerahandi: Exact Statistical Methods for Data Analysis.
West/Harrison: Bayesian Forecasting and Dynamic Models, 2nd edition.